Technical Communication

Second Edition

Technical Communication

Second Edition

David E. Fear

Valencia Community College

Scott, Foresman and Company

Glenview, Illinois

Dallas, Tex.   Oakland, N.J.   Palo Alto, Cal.

Tucker, Ga.   London, England

Photo Credits
All photos not credited are the property of Scott, Foresman.
20, Northwestern University; 26, Charles Gatewood; 56, Richard Stromberg; 61, Courtesy Illinois Bell; 99, Photo Researchers; 121, H. Armstrong Roberts; 177, 197, Joel Gordon © 1978; 246, Northwestern University; 286, Jean-Claude Lejeune; 317, Photo Researchers.

Library of Congress Cataloging in Publication Data

Fear, David E   1941–
    Technical communication.

    Includes index.
    1. Communication of technical information.
    2. Technical writing.   I. Title.
T10.5.F4   1981   808'.0666021   80-28991
ISBN 0-673-15401-7

An *Instructor's Manual* for *Technical Communication,* Second Edition, is available. It may be obtained through a Scott, Foresman representative or by writing to English Editor, College Division, Scott, Foresman and Company, 1900 East Lake Avenue, Glenview, Illinois 60025.

Preface

This second edition of *Technical Communication* retains the same fundamental goal as the first edition—to provide sufficient materials both on basic techniques and on a wide variety of applications to help students prepare for the writing and speaking situations they will encounter in the world of work. It also retains the same three-part structure, providing materials for mastering, applying, and sharpening communication skills. Thus, *Technical Communication,* Second Edition, has been designed with today's career requirements in mind, to help students learn to assess a communication situation, to write clearly and concisely, and to practice these skills by means of specific applications.

Part One, "Basics," examines in technical contexts the fundamentals of composition—planning, organizing, writing, revising. It especially encourages students to analyze the audience and the purpose of any communication before starting to write. In this revision, the actual composition process is treated more fully, with Chapters 5 and 6, on writing and revising, giving numerous concrete suggestions for building effective sentences and paragraphs and for improving diction. The planning worksheets have been simplified, and are accompanied by guidelines for using them. A chapter on selecting and constructing tables, graphs, and other illustrations rounds out this section.

Part Two, "Applications," applies the skills taught in Part One to specific writing and speaking situations. Considerable rearrangement and addition have been done here. Memorandums have been given their own chapter to better reflect their significance in business and industry. Chapter 14, "The Job Package," contains materials that help the student develop an effective resume and letter of application, some sound techniques for job interviews, and sample letters for accepting and rejecting job offers.

Part Three, the "Handbook for Reference and Review," remains essentially the same, including chapters on sentence structure, grammar, punctuation, and research techniques.

Two significant changes have been made throughout this edition of *Technical Communication.* Discussions and explanations have been shortened and tightened wherever possible, and the text is supplemented by more professional models and examples. Most of the previously used models have been either replaced or updated. Secondly, a wide variety of exercises has been added at the end of each chapter. Some exercises are intended for classroom discussion, but most are designed to provide written exercises of varying length and complexity. More exercises than any one instructor might want to assign are supplied, but I hope this will increase the text's flexibility in enabling instructors to better attain their own objectives with their students.

As an additional tool, an instructor's manual, which offers more suggestions for using and developing the materials in the text, accompanies *Technical Communication.*

Many individuals made major contributions to this edition. One battery of instructors made suggestions based on the first edition; another group criticized the revision plan; still others examined an initial revised draft. For this assistance, I would like to express my appreciation to the following people:

Julie Alexander, *Colorado School of Mines*
Ingrid Brunner, *Lehigh County Community College*
Don Richard Cox, *University of Tennessee*
Margot Haberhern, *Florida Institute of Technology*
Richard Harp, *University of Nevada*
Thomas Ireland, *Delgado College*
Theodore P. Johnson, *Iowa Western Community College*
Michael L. Keene, *Texas A & M University*
Jack Klug, *San Antonio College*
David M. Kvernes, *Southern Illinois University*
Wayne A. Losano, *University of Florida*
Joseph Lostracco, *Austin Community College*
Henry Dan Piper, *Southern Illinois University*
Robert S. Rudolph, *University of Toledo*
Freda Stohrer, *Old Dominion University*
Arthur Wagner, *Macomb County Community College*
James C. Work, *Colorado State University*

As usual, all of the Scott, Foresman staff did an outstanding job. Special thanks go to my wife, Linda, for typing and editing the manuscript and for developing many new models and exercises.

D. E. F.

Contents

PART THREE   HANDBOOK FOR REFERENCE AND REVIEW   321

Technical Communication

Second Edition

PART

7

ONE

Basics

1

Preliminary Considerations

Overview   *In order to prepare an effective piece of technical communication you must first decide what you are supposed to prepare, why you are preparing it, what restrictions there are on it, and when you must have it ready. Then you must analyze your reader. At the very least, you must determine whether the reader is highly technical (an expert on the subject), semitechnical, or nontechnical (a layperson).*

Technical communication is on-the-job communication designed to perform specific functions. It has been defined in many different ways, but seldom satisfactorily. Almost the only generalization that can be made about technical communication is that every instance of it is unique. Therefore, this text is designed to help you "zero in" on specific communication problems; it will advise you on composing the most important types of job-related communications and supply models, checklists, and guidelines. Seldom, however, will your communication situation fit exactly into the specific patterns and models given here. In order to help you meet any communication problem that may arise, this text will also discuss underlying principles and techniques of analysis that can be applied to any communication situation.

Understanding your situation is a key factor in becoming an effective technical communicator. To help you begin to analyze your situation, Chapters 1 and 2 introduce the *planning worksheet.* Before you begin filling one out, you must answer the important preliminary questions "What am I to prepare?" "Why am I preparing it?" "What restrictions are on it?" "When should it be ready?" and "Who will its readers be?"

What Am I to Prepare?

Before you can prepare anything, even a meal, you must know what is appropriate for the situation. You would no more wish to submit a handwritten memo when a formal letter is called for than you would a pot of chili when a soufflé is more appropriate.

Sometimes it is easy to determine what is called for. If you submit a formal progress report every Friday and it is now Friday, you know what to do. Often, though, you will have to think a bit. For example, suppose your boss, the chief therapist, asks you to prepare a report on Mrs. Gravely for Dr. Mordred. You know that a report is needed, but you must classify it a bit further. You check with the supervisor, review Mrs. Gravely's situation, and perhaps check your files. You find that Dr. Mordred is familiar with Mrs. Gravely's case but has not seen her in several days. You decide that neither her complete medical history nor a detailed account of her present hospitalization is needed. You decide to write a brief status report detailing Mrs. Gravely's present condition.

You might use a similar process in classifying an oral presentation. Suppose you are a graphic artist who has been asked to speak for about ten minutes to a group of high-school students who will be touring your office in a few days. Af-

Before submitting any kind of technical communication, you must understand the situation for which it is required. Usually, this kind of preliminary work starts by getting first-hand information from a supervisor or from someone in charge.

ter agreeing to the talk you realize that "a ten-minute talk to thirty seniors from Wilson High" is really all you know about your task. You therefore begin preparations by talking to your supervisor and to the adviser of the student group to learn more about the group's interests and background. With this information in hand you know what type of talk to prepare.

Why Am I Preparing It?

Determining exactly what you are supposed to prepare usually entails identifying why you are preparing it. This should be easy to do, but take the time to do it anyway. You may be preparing it because the boss asked, but why is it necessary in the first place? What will the boss do with it? If you are not certain, ask. Even if you are preparing the communication on your own initiative, take the time to clarify your purpose in your mind. Suppose, for instance, that you are a convenience store operator required to write a monthly report to the regional supervisor. A little thought tells you that the company has two purposes: to determine what products are and are not selling and to determine what stores are and are not producing. A bit more tells you that your real purpose is to report your month's activities in such a way as to show what a conscientious, effective manager you are. With this purpose clearly in mind you will probably prepare the report a bit differently than if you had just plunged in and begun writing.

What Restrictions Are on It?

What you are preparing to write or speak will usually be influenced by restrictions. Some restrictions are clearly stated as organizational policy; others are informal, unwritten laws, never really stated but never forgotten by the alert employee. Nearly every large organization does things its own way. If you change jobs and go to work for a competitor with the same clientele, you might find other communication restrictions. Some will appear trivial. One company might use modified block style for letters; another, full block style. (Letter styles are discussed in Chapter 8.) Honor such preferences.

More substantial restrictions may involve matters such as corporate image or philosophy, and they may drastically affect the types of communication prepared. These may concern not only the label you give items or the form you present them in but even what information you include and the tone you use. Failing to observe such restrictions could have serious consequences.

The restrictions you will encounter may be formal or informal, official or unofficial, written or oral. Check for written materials containing formal restrictions. Some companies publish style manuals that spell out restrictions for all formal communication. Others include them in such documents as job descriptions, standard operational procedures, and instructional manuals. To determine informal restrictions you will have to work a bit harder. Start by examining communications done by successful colleagues and supervisors. They will usually be glad to fill you in on unwritten restrictions or codes.

When Should It Be Ready?

One more question must be answered in determining what you are to prepare: "When is it due?" When a due date has been set, answering this question is as simple as it seems. Other times, however, the simplicity of this question is deceptive. Suppose no firm deadline has been set, but you know the reader will need the information by the first of the month. You have a chance to plan some strategy. When would be the best time to present your material? Should you hurry to get it in as soon as possible, several weeks before the reader needs it? Or should you wait until just before it is needed? Generally, you should try to get it to the reader as soon as you can without being rushed. Planning an early submission also gives you some leeway in case another important matter comes up. Receiving the material early gives the reader more time to consider and use your information. If it is well presented, submitting it early may even help you to make a good impression on readers before they receive work from others.

Occasionally—though not often—a later submission may be better strategy. If your readers receive the material too long before they need it, it may be forgotten, or at least not remembered as clearly and favorably as you would like. But do not wait until that last possible moment. Doing so not only risks the displeasure of the reader but will likely increase the critical attention he or she gives your presentation.

Planning your timing for submission also enables you to set up a tentative schedule for your work. Plan your time for organizing, researching, writing, and revising. Work through each phase carefully, methodically, and thoroughly, safe in the knowledge that, although you do not have time to waste, you do have time to do a good job.

Once you have determined what you are supposed to prepare, why you are preparing it, and what restrictions—including time restrictions—you are working under, you are ready to attack the next major stage in planning: analyzing the reader.

Who Will Its Readers Be?

Second in importance only to knowing what you are preparing is knowing whom you are preparing it for. In communicating with people, you will usually be trying to have a definite effect on them or to elicit a definite response from them. Hence, you will need to know what their reactions to your communication are likely to be.

All successful communicators have the knack of seeing things from the other person's point of view, of judging beforehand what effects their words will have on someone else. Those who fall short often make the mistake of assuming that what is clear to them is also clear to their reader or that what seems thorough to them is thorough enough for their reader.

However, what a reader sees may not be just what the writer intended to write, and what a listener hears is not necessarily what is said. You and a friend do not hear exactly the same thing when you sit listening to the same speaker. Your attitudes toward the subject, toward specific details and examples, toward the speaker's personality, company, or profession—all affect what you hear, as do your intelligence, your vocabulary, your knowledge of the subject matter, and even your ethical values.

Later we will examine techniques to design a presentation specifically for a given reader (or a given listener). First, we must concentrate on what to learn about that entity. Begin by identifying the reader. Is it an individual or a group? If a group, what kind? If an individual, is it someone you know? By name? By reputation? Or only by title?

Similar questions can be asked about individuals and groups. Say you have identified your audience as a small group. Now find out specifics. How well do the people in the group know the subject area? Are they laypersons? Authorities in the field? Or somewhere in between? Are they well-educated? What might their attitudes be toward the subject in question? Might they react strongly? Might they be biased? Do they have any strong preferences about method of presentation? Do they prefer a particular organizational pattern? Will they prefer complete detail or demand brevity? Do they like certain expressions or words? What else do you know about them?

Do not worry if you cannot answer all these questions. Few of us could, even about close associates. But the more information you can get the better. If you know the readers well, you should get a sizable body of information; but even if you merely have the names of some people halfway across the country, you can get some answers. Check any correspondence you might have had with them. Read anything you can find that they have written. Talk to colleagues who may have dealt with them. Even one or two bits of information may help you to prepare a more effective communication; for example, just learning that your readers demand brevity or that they have some particular like or dislike may save you a great deal of trouble.

Level of Technicality

You can almost always determine a reader's (or a listener's) level of technicality (knowledge of the general subject area) and prepare the communication at that level. A trained professional in the general subject area is a *highly technical reader;* one who has no specialized training or education in the area is a *nontechnical reader;* those in between are *semitechnical readers.* Place your audience into one of these three categories, and adjust your communication accordingly. Such general categorization is not as good as being able to write or speak directly to your reader, but it's often the best that you can do.

Highly Technical ● Think of a highly technical reader as knowing as much about your subject as you do but merely not having the specific information you are preparing to communicate. This reader may have the same job as you or may be your immediate supervisor or subordinate. He or she knows the standard terminology, symbols, and abbreviations and is familiar with the basic concepts and common procedures and apparatus.

The highly technical reader is rather easy to write for. You need not define any technical terms or provide detailed explanations of processes or intermediate steps. You can also name equipment or apparatus rather than describing it. Your only real worries with a highly technical reader are to be precise and to be clear. You cannot fool such readers, and they will not be satisfied with vague generalities.

Semitechnical ● The semitechnical reader is sometimes harder to identify than the highly technical one. Writing for these readers is a bit more challenging. They are familiar with your general subject area but are not specialists in it. They may work in a closely related field or in a different specialty within the same one. Such a reader may be a supervisor over your department and several others, with only a broad, general knowledge of each. He or she may be a beginner in your field or a client or customer paying for your special knowledge.

Since the semitechnical reader has only a limited understanding of your field, you will have to explain unusual and highly technical processes and apparatus. He or she will know many basic technical terms, but the more advanced ones will need clarification. Often a few working definitions in parentheses will fill the gap; at other times you can paraphrase to avoid highly technical terms. Some sophisticated concepts can be simplified; others can just be avoided.

Nontechnical ● The nontechnical reader, or layperson, probably has no background in the subject area but may know something about it. Depending on the nature of the subject matter, you might expect the average well-informed person to know a bit about it or—if yours is a new or little-publicized field—nothing at all about it. You must try to determine the extent of this knowledge. And that is often just the beginning of the difficulty.

On the one hand, you must keep in mind that your reader has no training in your field. This means that you must simplify everything. You must define many terms and avoid using highly technical concepts. You must explain processes and describe apparatus. On the other hand, you must assume that your reader can follow normal reasoning patterns, has a normally good vocabulary, and is generally well informed. Ineffective nontechnical writing—and there is much of it—is about equally divided between that which is too technical and that which is too simple. If you must err, be too simple, but try not to belabor the obvious.

Before submitting a piece of nontechnical communication, try it out on a handy nontechnical reader. Such a person can often spot problems that would never occur to you. You know the material completely; you know what you want to communicate. But matters clear to you may be identified by your non-technical proofreader as needing definition or further explanation. He or she can also spot places where you have oversimplified.

Combination Levels ● If you have only one reader and that person is clearly highly technical or clearly nontechnical, you have few worries about level of technicality. Unfortunately, this is rarely the case. You will sometimes write to groups whose members have varying levels of technicality. Even with just one reader the best you can do at times is to estimate the level of technicality as "fairly high" or "pretty low."

For these ambiguous situations, remember that writing below the appropriate level of technicality will bore the reader and waste time but that too high a level will be almost impossible to understand. When in doubt, the lower of two levels is better. This is always true for an individual reader and is almost always true for a group. Write at the higher level only when you identify a primary reader (the reader you are most concerned with impressing or satisfying) whose level is higher than that of the secondary readers. Write directly for the primary reader; do not worry about the others.

Some documents are actually written at several levels of technicality. In proposals, for instance (see Chapter 11), you usually explain matters nontechnically first, then go through them highly technically in more detail. Long, formal reports (see Chapter 10) generally include a highly technical main text, supplemented by front matter and appendixes at a lower level. Other times, highly technical appendixes supplement a semitechnical or nontechnical main text. In either case, the main text is written for the primary reader.

Determining the Appropriate Level ● If you do not know your reader personally, begin by considering his or her position. What is the job's title? What training or education is generally required for the job? If the reader is a manager or supervisor, do people in similar positions generally have technical or administrative backgrounds?

A little discreet checking will usually help you to determine the level of a person within your own organization. If your reader is outside your organization, check with people holding similar positions in your organization or ones similar to yours. This takes a bit of effort, but writing an important document at the appropriate level should make the effort worthwhile.

The following guidelines may help:

1. The farther someone is from you on your company's organizational chart—whether above you, below you, or in some other department—the lower that person's probable level of technicality in your subject.
2. A supervisor over one department is usually highly technical in most work done by that department. A supervisor over several departments is probably semitechnical but may be highly technical in one area. Usually, the more different specialties supervised, the less the technical knowledge in individual areas.
3. Members of your department are usually highly technical, others within the organization semitechnical, and top management and outsiders nontechnical.
4. People in other organizations are probably of the same technicality as those in similar positions within your organization.

Starting the Planning Worksheet

As you answer the preliminary questions and analyze your reader, jot down your answers on a planning worksheet such as the one in Figure 1.1. A worksheet is just an organized form for recording notes to yourself. Once you have become experienced, you may be able to make your notes more casually or even keep them in your head, but use the planning worksheet for now. The following simulation shows how the first section, "What, When, and Why," might be completed.

What, When, and Why? _____

_____

_____

_____

_____

_____

Who Is My Reader? _____

_____

_____

_____

_____

What Is My Thesis or Central Purpose? _____

_____

_____

_____

_____

_____

What Approach Will I Take? _____

_____

_____

_____

_____

Figure 1.1 Planning Worksheet

What, When, and Why?   SOP for NCR 3600 keypunch. Follow format

used by billing and credit departments. Should be ready within

two and one-half weeks (October 24) for testing. Ready for Mr.

Acevedo by November 4.

_____

_____

_____

_____

Who Is My Reader?   Hector Acevedo, hospital controller, and new

keypunch operators. Should aim at clarity for operators. Semi-

technical but with no special knowledge of this machine. Can use

standard terminology and keypunching.

_____

_____

What Is My Thesis or Central Purpose? _____

_____

_____

_____

_____

_____

What Approach Will I Take? _____

_____

_____

_____

_____

_____

Figure 1.2 Planning Worksheet

As the director of data processing for a large hospital, you have just been asked by your new boss, hospital controller Hector Acevedo, to prepare standard hospital procedures for several machines in your department. He has asked that they be ready within one month. Answering "What?" is no problem: you will be preparing standard operational procedures for each of five pieces of equipment.

"Why?" is a bit trickier. Sure, you were asked to by the new boss, but the real purpose will be to assist inexperienced operators in using the equipment. "When?" seems at first glance to be one month, but then you realize that you do not really have that long. It will take at least a week to let some of the operators try out the new procedures so that you can revise them before you make your final copies. Furthermore, the new controller is such a stickler for promptness that you want to get copies to him at least two or three days earlier than required. Next you consider possible company restrictions. You know that some departments use identical layouts for their SOP's, and you decide to follow suit. You can now complete the "What, When, and Why?" section of your planning worksheet. (See Figure 1.2.)

Now turn to reader analysis. Determining level of technicality is simple enough: both Mr. Acevedo and the new operators who will be using the machines are semitechnical. None of the operators are experienced on these machines, but they have been trained on similar ones. Acevedo has been around data processing all of his career. Since Acevedo strikes you as a "no-nonsense" type, interested in results, you decide to concentrate on making the procedures clear to the operators. You know that they are high school graduates, some with a bit of college or vocational training. Their main concern will be clarity. You can now complete the "Who is my Reader?" section of the worksheet.

• CASE STUDY

Study the following communication situation and complete the "What, When, and Why?" and "Who is My Reader?" sections for a planning worksheet.

You are manager of a men's clothing store, the local Parkwood branch of a regional chain called "The Dude." Just a few miles away a large shopping center is being built. Several men's stores, including a new branch of "The Dude," will be located in the new center. You will be made manager of the new, much larger store—at a new, much larger salary, of course. Tom Gastineau, regional manager, has asked you to submit suggestions for best utilizing your old store. Gastineau was your first manager when you started in retailing and is still a good friend. A stickler for punctuality, he also likes things done formally and precisely. He seems concerned about the two stores competing with each other.

• EXERCISES

1.   Interview one of your major field instructors or, even better, a practicing professional in the field you wish to enter. Find out what kinds of writing that person does and the types of readers for each. Prepare a brief list classifying these readers as highly technical, semitechnical, or nontechnical.   MODES, APPLICATION
2.   Select a difficult passage from one of your major field textbooks and rewrite it for a classmate with a different major. Swap finished products to determine if your level of technicality was low enough.

2

Planning the Presentation

Formulating a Thesis

Selecting an Approach

    Point of View

    Tone

    Mode

    Scope

Completing the Planning Worksheet

Case Study

Collecting Evidence

    Types of Evidence

    Examples

Checking Your Logic

    From Evidence to Recommendations

    Some Fallacies to Avoid

Exercises

Overview   A thesis states the crux of an entire presentation in one pithy sentence. After developing a good thesis, select a point of view, a tone, and a mode; then determine your scope. Use the thesis to collect appropriate evidence—empirical evidence, printed or filmed material, and testimony from others. Use examples to vivify your evidence and to introduce it. In organizing your evidence, guard against the hasty generalization and fallacies such as the post hoc fallacy, the faulty dilemma, begging the question, and red herrings.

The more thoroughly you plan a piece of written or spoken communication, the more quickly you will be able to complete it. Contrary to what many people believe, the overall composing time is less when the writer spends more of it on advance planning. The more thoroughly you plan, the faster the first draft will go, the fewer major revisions you will need, and the less likely you will be to tear up a draft and start all over. It might be painful at first, but give the suggestions here a fair try. They will help.

Formulating a Thesis

A thesis states the crux of an entire presentation in one pithy sentence. Everything else is elaboration, clarification, justification, or exemplification of that one point. A good thesis does not merely identify the subject you are discussing ("I am going to discuss the advantages of turbocharging automobile engines."); it states the central point you are going to make about that subject: "Turbocharging a typical automobile engine will improve both its performance and its fuel economy."

Notice that this thesis (like all good thesis statements) performs three functions:

1. It identifies a clearly defined subject.
2. It makes clear your attitude toward that subject.
3. It focuses your reader's attention on the specific aspects of the subject that you will discuss.

The sentence above, for instance, identifies your limited subject as advantages of turbocharging ordinary automobile engines. It clearly communicates your positive attitude toward the subject. Finally, it focuses your reader's attention on the specific points you will discuss, improved performance and improved fuel economy. The ineffective thesis statements below fail to perform at least one of the three functions.

Geophysics is a good field today.
Plants grown from bulbs, such as tulips, are difficult to grow in warm climates.
It is important for the detective to protect the crime scene.

Notice how these improved theses do perform all three functions:

> Because of the current emphasis on alternative sources of energy, geophysics offers exciting and lucrative job opportunities.
> Because they need long cold seasons for their dormant periods, it is not feasible to grow tulips and similar plants in Florida or south Texas.
> The detective's first responsibility at the crime scene is to protect it until all possible evidence can be located and processed.

By formulating your thesis early, you can more easily and effectively determine the exact approach you wish to take (discussed in the next section) and you can more efficiently determine what evidence to include.

Do not be surprised if you find yourself changing your thesis as you accumulate more evidence. This often happens. Certainly you would rather present a thesis that accurately reflects the actual evidence than twist the evidence to fit a preconceived thesis. In gathering evidence to support your thesis about raising tulips, for example, you might find that the evidence indicates that it is really not all that difficult to raise tulips in the extreme South. So you change your thesis to something like this: *Contrary to popular belief, a few simple precautions can make raising tulips in Florida or south Texas rather easy.*

Below are sample thesis statements for some of the common types of technical writing discussed in Chapters 3 and 4.

1. *Technical description:* The Carnegie pest-trap has three main sections: a closing mechanism, a holding tank, and a wire gate.

2. *Technical definition:* CPR (Cardiopulmonary resuscitation) is a technique for maintaining or restoring heart and lung function.

3. *Comparison-contrast:* The new panoral dental X rays offer advantages over the traditional bitewing X rays for the dentist, the technician, and the patient.

4. *Narrative:* My business activities on the Paducah trip consisted of three distinct phases: First, I was shown the new machines in operation; next, Georgia Radsdale showed me production figures; finally, Jack Mellhorn and several foremen discussed minor difficulties with the new machine.

5. *Process analysis:* The traditional watch is powered by a mainspring that unwinds at a fixed rate.

6. *Instructions:* To get the smoothest possible finish, just follow the three major steps carefully and keep your steel wool wet.

7. *Classification:* Your prospects will fall into three types: those doing a small amount of business with us, those doing business with our competitors, and those not realizing that they need our merchandise.

Remember, your thesis sums up the entire presentation in one sentence. Even if your reader were to forget everything else except that one sentence, your time would not have been completely wasted.

Selecting an Approach

Your final piece of preliminary planning is selecting the most effective approach. Four specific elements are involved: point of view, tone, mode, and scope.

Point of View

The three basic points of view are the first person (designated by *I, we*), the second person (*you*), and the third person (*he, she, it, they*). Each is useful in certain types of technical communication. Additionally, each of the three can be presented in either active or passive *voice*. Active voice (he tightened the screw) is more common in most communication, but many technical communicators use the passive (the screw was tightened). The guidelines below will help you to determine which point of view and voice to use.

1. Use first-person, active voice sentences to offer conclusions and to assume responsibility:

I would prefer the electronic to the electromechanical.
We entered the computer room at exactly 3:15.

2. Use first-person, active voice sentences in the narrative section of reports:

I began by performing an SMA 12.

3. Use a second-person point of view to write instructions:

Next, sand the board with 300 grit sandpaper.
You should then let it dry overnight.

(Notice in the first example that *you* is implied without being stated.)

4. Use a second-person point of view to establish a direct one-to-one relationship and to show understanding for the other person's position:

You have no doubt noticed that most of this text is written in second person.
You can probably relate to it better than to a more detached third person.

5. Use third-person, active voice sentences to show a thing acting upon a thing or a person performing an action:

The fork then slides across the conveyer.
He picks up at least three pallets on each trip.

6. Use the passive voice to emphasize the receiver rather than the doer of the action.

The windshield was totally destroyed by the limb.

7. Guard against ambiguous passive constructions:

The bolt is attached. (Is this an instruction telling someone to attach the bolt, or is it a reminder that the bolt is already attached?)
A new set of procedures has been prepared. (Sentences such as this often fail to indicate who performed the action.)

8. Be careful of overusing *I*. "I this, I that, I something else" can become tedious. An occasional well-chosen passive or a smooth shift to another point of view is sometimes helpful.

9. Use the passive sparingly. Overuse of the passive is one of the most common criticisms of technical writers.

10. Remember that third-person, active voice sentences are the most commonly used in technical writing.

Tone

Tone is your attitude toward your subject and your reader as it shows through to the reader. Almost any common adjective that describes a person's attitude could describe the tone of a piece of technical communication. While point of view helps to set the tone, any point of view can be written to sound serious, businesslike, and formal or casual, friendly, and informal.

Tone is created through several means. Your selection of details, your choice of words, even your organizational pattern affect the tone. The most important thing is to keep your intended tone in mind: remember that you want to come across in a particular way. Obviously, the easiest tone to develop is the one that you really feel; however, you have no doubt learned by now that doing that is not always wise. Sometimes you must make a certain letter sound "tough" even though you do not feel very tough. Or you must make a memorandum to a disliked, dictatorial supervisor sound courteous and respectful even though you would prefer to call him or her names.

Let's examine the two most common tones in technical communications:

1. *Strictly business:* Most reports, proposals, job descriptions and operational procedures—and some letters—are composed in this tone. Point of view is usually (but not necessarily) third person. The passive may be used a bit. Vocabulary is standard, with no slang or colloquialisms. Technical terms are used liberally if needed. Organization is straightforward and efficient. Relevant information is given in sufficient detail, but no extras are thrown in. This tone works well with strangers, executives, supervisors, and large, heterogeneous groups. The model reports in Chapter 10 are good examples of the strictly business style.

2. *Businesslike but relaxed:* This tone is used in communications that must attend to business but that aim at creating relationships slightly more personal and warmer than a strictly business communication would achieve. It also works well in speeches to small groups and in memos and reports to people you know well. To create this tone, use whatever point of view you

The tone of your technical communication affects the way in which your presentation is received. Checking your organization's point of view on an issue ensures selecting an appropriate tone.

need. (First and second are more common than third.) Use a good, proper vocabulary, but don't be afraid to use a well-chosen colloquial or even slang expression. Keep the organization functional, but digress a bit if you have a reason to. In short, relax, and your tone will too.

Now let's look at two tones to avoid:

1. *Contentious:* A contentious tone makes you "come on too strong," sounding pushy and self-centered. Overuse of *I* or *we* is a primary cause. Be especially careful of sounding contentious in letters and memos. ("Your mistake cost us money.")
2. *Obsequious:* Nearly the opposite of contentious, an obsequious tone carries respect, courtesy, and humility too far. While lack of respect for your reader is conveyed by contentiousness, obsequiousness conveys a lack of respect for yourself. Be careful of it in letters of application and in other situations in which you are trying hard to be polite and courteous. ("Apologizing in advance, your faithful servant. . . .")

Here are some guidelines to help you to choose the most effective tone:

1. Be careful of carrying the businesslike-but-relaxed tone too far, especially with those you do not know well. Some people resent strangers trying to become too familiar.

2.   Guard against a too-relaxed tone when you are unsure of your reader's level of technicality. Writing in a casual tone below your reader's level can easily sound condescending.

3.   Double-check your tone when you feel very strongly about your reader or subject. Put your first draft away for a while to freshen your perspective before you check it over.

4.   Expect to expend a great deal of effort achieving a tone that runs totally counter to your real feelings or your basic personality. Some of us can sound tough and angry more easily than others, just as some of us have difficulty sounding mild and unassuming. When you face such a situation, check your draft line by line.

Mode

Your mode of communication is your means or manner of delivery—speaking or writing. If you choose to speak, you can use the telephone or the tape recorder or you can deliver your message in person. If you deliver it in person, you can speak from notes or read from a script. If you write, you can choose between, say, a memorandum or a letter. If you choose the memo, you can send a formal, typed memorandum or a brief, handwritten note.

Unfortunately, there is no set of rules or procedures to tell you definitely what mode to use in what circumstances. Some situations clearly dictate a certain mode; others do not, and any of several modes may work effectively.

Although your choice of a mode must depend upon your analysis of the reader and the situation and upon your common sense, you should find the following basic guidelines helpful.

Written Versus Oral

1.   Use a written mode when you need a record of the communication.

2.   Use a written mode when you are presenting a large number of statistics or hard-to-follow procedures.

3.   Use an oral mode when you wish to answer questions or clear up possible ambiguities concerning your presentation.

4.   Consider using both written and oral modes for important, complicated presentations: Submit a written, formal document with an in-person, oral explanation.

Formal Versus Informal

1.   Use a relatively formal mode—report, letter, formal prepared speech—for most executives, especially unfamiliar ones.

2.   Use a relatively formal mode for documents that will be official or that will be used for formal reference.

3. Use an informal mode to establish a relaxed, friendly relationship.

4. Use the more formal modes to address strangers or in any situation in which you are unsure.

Scope

Your *scope* is the amount of material you present, both in breadth and depth. Your thesis is the main determinant of breadth, since your presentation must deliver what the thesis promises. Even so, you will have to determine the depth—the amount of supporting or clarifying evidence to include. The following guidelines will help you.

1. Avoid presenting broad subjects shallowly unless the situation specifically calls for doing so. It is usually preferable to narrow your subject and cover it in more detail.

2. Formal modes require broader scope, especially more background information and more thorough introductions.

3. Written modes can have a wider scope than oral ones. The reader can study and analyze a document; a listener must comprehend more quickly.

4. Make the scope broad for any document intended for future reference. In three years or even three weeks the reader may have forgotten much of the background information that is obvious now.

5. If you are undecided whether to include certain information, include it. The reader or listener will be more bothered by too little information than by too much.

Completing the Planning Worksheet

Once you have determined a thesis and chosen your approach, record your results on a planning worksheet. The following simulation shows how you might complete one.

> Suppose you are the supervisor of the Bed and Bath department of a large department store. You have just realized that what you thought would be a hot item—a hand mirror in a large conch shell—is not moving. You know that this item is a personal favorite of Martha Dusek, head specialty buyer and assistant manager. So you know immediately that she is the one to notify. You also know that a typed memo is the best mode; there is no need for anything more formal, but she does like things in writing. You realize that you should act as soon as possible because the item will probably sell better in the new Surf and Sun department being readied for opening soon. A bit more thought, and you determine your thesis: Item JB I673 F, conch shell mirrors, should be moved from Bed and Bath into Sun and Surf, where they will fit in and sell better. You decide on a relaxed, first-person approach with a very limited scope. Martha is a good friend, and she knows the background. So you complete a planning worksheet as shown in Figure 2.1.

What, When, and Why?   Memo to Martha Dusek; as soon as possible;
no restrictions; for her immediate action.

_____

_____

_____

Who Is My Reader?   Martha Dusek, head buyer and assistant manager,
with copy to Brad Lochner, supervisor of new Sun and Surf department. Highly technical. No special problems.

_____

_____

What Is My Thesis or Central Purpose?   Item JB 1673 F, conch
shell mirrors, should be moved from Bed and Bath into Sun and Surf
where they will fit in and sell better.

_____

_____

What Approach Will I Take?   Relaxed; first person; concise with
limited scope; typed memo.

_____

_____

_____

Figure 2.2 Planning Worksheet

• CASE STUDY

Study the following situation. Do the necessary preliminary study, making up any information you need, and complete a planning worksheet for a communication to fit the situation.

As the new manager of a 450-unit apartment complex, you have been swamped with complaints about problems with the master television antenna system. After talking with local cable officials you determine that for $20,000 you can have cable installations put into every unit. You would then be relieved of all maintenance responsibilities for the system, since your tenants would deal directly with the cable company. You know that the apartment complex owners, in Fort Worth, Texas, several hundred miles away, are willing to spend money if they are convinced that the expenditure is justified. You also know that they want all possible information before making any decisions. It is now late October, and they might want to spend the money this year, for tax purposes. All of your relations with them have been completely businesslike.

Collecting Evidence

As mentioned earlier, one reason for developing a thesis early is to help you select supporting information. Having a sharp, clear thesis will enable you to sort through possible evidence, keeping only that which you can actually use in your presentation. Do not waste time gathering evidence that is vaguely related to your subject but that does not fit your thesis and that you will not want to include in your presentation. *Step one in collecting evidence is to study your thesis to determine exactly what kind of supporting evidence is needed.* And don't forget to consider your scope and your intended reader.

Assume, for instance, that you are developing the following thesis: "Reporting suspected cases of child abuse can save lives by protecting the child from further abuse and by forcing offending parents to get the help they need." The thesis suggests that your emphasis will not be so much on showing the magnitude of child abuse but rather on showing the procedures followed in reporting abuse and in caring for abused children and abusing parents.

Types of Evidence

Empirical evidence—evidence that can be verified with the senses—is the most reliable and the most difficult to dispute. Look first for cold, hard facts that can be verified by counting, measuring, seeing, or touching. And collect your data firsthand if at all possible; firsthand evidence clearly presented carries a great deal of weight. Unfortunately, you will often have to collect your information from sources other than your own personal experience, going to printed material or other people.

Searching the Literature ● Having collected all the firsthand information you can, see what you can locate in books, periodicals, and other printed sources. Your best starting place is usually the subject card section of the library's central card catalog. You will find the *Reader's Guide to Periodical Literature* helpful in locating potentially useful articles from popular magazines. Check the list of periodical and reference guides in Chapter 18 for a guide that can help you locate articles in more specialized journals.

As you skim the articles and look through tables of contents and indexes of the books, keep your thesis in mind to prevent wasting time on material you will not be able to use. As you locate information supporting or developing your thesis, take good, thorough notes and keep complete bibliographical information on each source (check Chapter 18 for complete details).

When you run into contradictory information, use the following guidelines:

1.   More recent information is generally more useful.
2.   Give more weight to articles in professional journals (usually published and edited by leading people in the field) than to those in popular, mass-oriented magazines.
3.   Check the credentials of the author or of supposed authorities cited or quoted by the author. Give the most weight to those with solid qualifications in the relevant field. Do not be fooled by fame. Someone famous in one field need not be authoritative in another.
4.   Be sure to paraphrase or quote accurately. Be certain that the authorities you cite actually say what you claim they do. Do not take items out of context unless you are certain that the context supports your interpretation.
5.   Do not use an opinion held by the minority of experts in the field unless you make clear to your reader exactly what you are doing. Be especially wary of the one prominent authority whose opinion differs from the majority of those in the field.

Getting Information from Other People ● Frequently, you will be looking for information that is too recent to have been published or information of a kind that does not generally get published. You do not have it close at hand, but you know of other people who might have it. You can use one of three means to get that information: interviews, letters of inquiry, or questionnaires. Whichever method you choose, take it seriously; prepare carefully for your interviews and prepare your questionnaires or inquiry letters equally carefully. You will probably be dealing with busy people, imposing on them. A poorly conducted interview or a poorly presented inquiry simply will not get you the information you need.

Letters of inquiry are explained in depth in Chapter 8 and job interviews in Chapter 14, but here are a few suggestions about questionnaires:

Collecting evidence for a presentation may mean constructing a brief questionnaire and conducting personal interviews.

1. Keep it as short as possible. A busy person will be more likely to answer a shorter questionnaire than a longer one.
2. Double-check each question to make it as clear as possible. Ambiguous or vague questions may not be answered. Even worse, they may be answered according to a totally different interpretation than you intended, giving invalid results.
3. Use questions that can be answered as easily as possible. Use essay questions—where the reader must compose a response—only as a last resort. Instead try techniques such as dichotomy questions—where the respondent chooses between true or false, yes or no, and the like—and multiple choice questions. Also useful are checklists or ranking lists, where the reader ranks a list of items in order of preference. Fill-in-the-blank questions are a bit more challenging to answer and to compose, but they are easier to work with than essay questions.
4. Leave space for comments or other unsolicited responses.
5. Make certain that the level of technicality fits your intended respondents.

6. Prepare a letter of transmittal to introduce your questionnaire, to explain your need for the information and your credentials, and to express your appreciation for the respondent's time. Chapter 8 will be helpful here.

7. Include a stamped, self-addressed envelope.

Examples

Examples prove very little, but they help to make almost any point clearer and more believable. Most effective communicators first explain their points with concrete, factual details, then clarify and vivify them with examples. For instance, if you were developing the thesis previously mentioned about reporting child abuse you might follow an actual case from the physician reporting suspected abuse through the final dispensation. Or you might use a negative example, showing what happened in a case where a physician did not report suspected abuse.

Examples also make excellent introductory devices. If you are looking for a way to get the reader's interest and focus it on your thesis, try starting with your most vivid example. You might begin the presentation on reporting suspected child abuse with the negative example just mentioned. You would certainly get almost any reader's attention and could then easily focus that attention on your thesis.

Checking Your Logic

In putting your evidence together, you will arrive at generalizations; in fact, your thesis will often state the generalization, followed by specific supporting evidence. *Generalizations* are the bases of inductive reasoning and the scientific method; they make inferences about all members of a group from observation of particular members of the group. You observe three automobiles getting improved mileage when driven at or below fifty-five miles per hour; you may generalize that other cars will get the same results. Unfortunately, generalizations are often unreliable, and the *hasty generalization* is one of the most common logical fallacies. A hasty generalization is exactly what its name implies—a generalization based on too little or unreliable evidence.

Obviously, then, although generalizations are necessary, they must be made carefully. Suppose, for instance, you were looking for evidence to support your thesis that the citizens of your community feel strongly against raising taxes to add deputies to your local sheriff's department. Sure enough, you distribute questionnaires to three hundred citizens picked randomly at a local shopping center and get results 80–20 in your favor. You present your recommendation to the county administrator, who rejects it; and later you are shocked to see the referendum pass handily. Where did you go wrong? To begin with, three hundred out of many thousands of voters is a very small sampling. Furthermore, your random subjects at the shopping center (which happens to have been the most exclusive in your area) do not necessarily represent typical voters. Study the suggestions for making generalizations below:

1. Remember that generalizations are really just predictions. If something is true four times it will likely be true a fifth, but not necessarily. The more particular instances you base a generalization on, the sounder it becomes. If you are the least bit doubtful, run a few more tests, question a few more people. Then express your conclusions as a likelihood, a probability, not a certainty.

2. Use as much evidence to support your generalization as you can. Your conclusion that gasoline mileage is better at below fifty-five than above it is more impressive when you add some technical details about engine efficiency.

3. Make certain that the specific cases upon which you base a generalization accurately represent the entire group, especially if you are generalizing about attitudes or opinions. If you cannot greatly increase the size of your sampling, make certain that the people you do question are representative of the entire group. In the example of the tax referendum, you should have contacted citizens of all financial levels—not just the most affluent.

One final word of caution: Beware of the self-fulfilling prophecy, of finding what you are looking for. If you begin with a broad generalization as a thesis and seek out particulars to support it, you can easily overlook contradictory evidence. Dig out your evidence objectively, and do not hesitate to change your thesis if the evidence warrants it.

From Evidence to Recommendations

The ultimate purpose of much job-related writing is to offer recommendations. Even an ordinary accident report is written largely to recommend corrective and preventive measures. In this, and in any piece of writing offering recommendations, you must carefully lead up to the recommendation, showing your reasoning along the way, allowing your reader to check up on you.

To see how this is done, let's backtrack from a typical recommendation, much as a reader might. Assume that you are recommending the purchase of six Brand *X* typewriters for your department. Your boss has asked you to try out and study several competitive brands to determine which she should buy.

Recommendation: We should purchase six Brand *X*, Model 627V machines from Nesbitt distributors.

Final conclusion: Brand *X*, Model 627V is the best buy to meet our needs.

Preliminary conclusions:

1. Brand *X,* Model 627V, carried by Nesbitt Distributors, has the lowest initial cost.
2. Brand *X,* Model 627V and Brand *Z,* Model *B* have essentially the same maintenance and repair costs.
3. Nesbitt Distributors offers the only unconditional guarantee.
4. Nesbitt, Jones and Jones, and Raystos are all well-established well-respected distributors.
5. All distributors can deliver promptly.

Findings:

Cost per Machine (in lots of six)	Guarantee (type and length)	Delivery Time (in days)	Yearly Maintenance (in dollars per machine)
Brand *Y*—$673.27	One year—parts	30	$23
Brand *X*— 591.29	One year—unconditional	21	24
Brand *Z*— 670.18	One year—parts	21	31

Brand *Y* is distributed by Jones and Jones, which is very highly thought of; Brand *X* by Nesbitt, also completely reputable; Brand *Z* by Raystos, the area's oldest supply house, also completely reputable.

Your reader need not agree with your recommendations, but she should be able to determine how you reached them. If she disagrees, she can see where in your chain of reasoning she differs with you.

Some Fallacies to Avoid

When putting together your evidence, you must be especially careful to watch for certain types of fallacious reasoning. The following fallacies occur all too often.

1. *Post hoc* is the common designation for the fallacy officially termed *post hoc ergo propter hoc*—"after this therefore because of this." In simpler terms, the post hoc fallacy concludes that when two events occur, one right after the other, the first causes the second. Obviously, this is not necessarily so. A third event may have caused both; what appears to be the cause may actually be the result; the events may have no relationship at all. Post hoc reasoning is the cause of many superstitions: A person who falls soon after walking under a ladder may blame the ladder rather than clumsiness.

To avoid the post hoc fallacy, double-check all conclusions involving cause and effect. Demand more than just coincidence before you determine that one event caused another.

2.   *Faulty dilemma,* also called the *either-or fallacy* or the *black-white fallacy,* assumes that only two alternatives exist, that something is either black or white with no grays possible. Once in a great while it is true that we have two and only two choices, but generally other choices or compromises are available. Most situations are far too complicated for an "all or nothing at all" type of reasoning. Few questions have simple answers; few experiments are total successes or total failures.

To avoid the faulty dilemma, question all either-or situations to make sure that there are indeed only the two choices. Most of the time you will find that there are others.

3.   *Begging the question* involves no begging or pleading; rather, it involves an assertion that is made and assumed true without sufficient proof. Notice the hidden assertions in the following: "We citizens must do something about this unfair collusion." "Why do we tolerate such illegal behavior?"

To avoid begging the question, check your conclusions carefully for underlying assumptions. Proving your ultimate conclusion does no good if it is based on other conclusions that are not proven. Take nothing for granted that your reader might question. Just because you know that something is true, do not assume that others will accept it without evidence.

4.   *Ad hominem and other red herrings* are related fallacies. A *red herring* is an argument or a piece of evidence that is irrelevant to the conclusion in question. *Ad hominem* literally means "to the man"; it describes arguments revolving around the holder of an idea rather than the idea itself, and it is the most common form of red herring. For instance, the argument that you should purchase equipment from Manufacturer *X* because of that firm's sales representative is ad hominem. A sales representative's charm, honesty, or political position does not make his or her *product* superior, and even the least desirable salesperson may have a good product. Guard against all forms of red herrings. Examine all your evidence carefully and discard that which is not relevant.

● **EXERCISES**

1.   Study the sentences below to identify which, if any, of the following fallacies they contain: hasty generalization, post hoc fallacy, faulty dilemma, red herring, begging the question.

   a.   Since we've had our new system in operation, we have handled 20 percent more cases. Obviously, it's worth what we paid for it.
   b.   Please accept my proposal. I've never given you a bum steer yet. You can certainly trust my judgment.
   c.   If we do not increase sales, we will have to close our doors.

d. Why do we tolerate such dishonest practices?

e. Welsh must really be loaded; he's an M.D.

f. If we don't get more oil from the Arabs we won't have enough.

g. Women work better than men on dull, repetitive tasks.

h. The suspect was known to hang out in all three of the bars that were robbed; he must be guilty.

i. Only a strong local union can stop these unfair labor practices.

j. You can tell that the JB–I37 is a great machine. Just look at the attractive design.

k. How can we stop this illegal price gouging?

l. Both of my former women bosses were excellent; I'd rather work for a woman anytime.

m. If we can hold the line on prices, we are sure to increase our sales.

n. McGeorge has a large family and is a veteran, so he'll get my vote.

o. Jackson's investments haven't even kept pace with inflation; let's put our money somewhere else.

p. Unless we stand firm, they will run right over us.

q. The Alsopp salesperson is so nice; we really should buy more of their products.

r. We've had nothing but bad luck ever since we went to a word processing system; the old way is obviously better.

s. Why don't we consumers rebel against these ripoffs?

t. He's a typical PR man; you can't believe anything he says.

2. Assume that your home town has just been given a large plot of land for use as a park. Prepare a questionnaire to find out from area residents what sorts of facilities they want in the new park.

3. Analyze the theses below and rewrite those you consider ineffective:

a. There are four classes of fires.

b. Diabetes mellitus is a disease of the pancreas.

c. To apply panelling to a sheetrock wall, just glue it on.

d. A timing light is easy to use.

e. Breeding Bettas (Siamese fighting fish) is fun and profitable.

f. Public relations involves much more than just advertising.

g. This paper will discuss the relative costs of shale oil, coal slurry, and gasohol.

h. An outside filter has four main sections: a motor, an intake tube, a filter box, and a return tube.

i. Spouse abuse is a major social problem.

j. The new NCR terminals are better than those we use now.

4. Select an article from each of two journals in your major or some other field of interest. Study each carefully, then complete planning worksheets you think the authors might have used to help develop the articles.

5. Write a thesis for a paper to be presented to a local governmental agency— county commission, city council, or the like—suggesting specific facilities for inclusion in the new park mentioned in Question 3.

6. Prepare a planning worksheet for the paper in Question 5.

3

Five Technical Modes

Description

    Photographic Details

    Impressionistic Details

    Photographic vs. Impressionistic Language

    Organizing the Description

Narration

    Basic Narrative Patterns

    Selecting Narrative Details

    The Language of a Narrative

Definition

    Working Definitions

    General Definitions

Comparison and Contrast

Partition and Classification

Exercises

Overview   *The five organizational patterns discussed in this chapter will help you build a good, clear technical communication.* Description *and* narration *may be either photographic or impressionistic, depending on the choice of details. You may need to include* definitions *of terms; working definitions give a brief translation of the terms, while general definitions give a fuller understanding.* Comparisons and contrasts *may be organized in either point-point or one-other style.* Analysis by partition and classification *can be used to divide a complex subject into workable parts or classes.*

As you saw in Chapter 1, written technical communication varies from a few lines scribbled in longhand by one person to multi-volume printed works written or edited by entire departments. Similarly, oral presentations may vary from a few words to a few hours of words.

Fortunately, nearly all written and oral technical communication can be developed through the use of a few basic patterns. For some communications you will need only one simple pattern; others may require a combination of several. A typical formal report, for instance, may require all of the patterns discussed in this and the next chapter.

Short models illustrate each of these patterns, and you will be asked to construct a sample of each. After you have learned to develop good short papers of each type, longer papers should present no problems.

Description

From age one you have been describing things—people, objects, pets, all sorts of things. When you developed the power of discrimination that let you recognize one person as "Daddy" rather than calling all males by that term, you were *describing in your mind.* You were selecting details—perhaps a combination of distinguishing characteristics such as thick, brown, wavy hair, a large nose, a deep voice—that set "Daddy" apart from other men. Just so, in technical communication, you will be sorting out and selecting details that distinguish and differentiate most clearly.

Almost certainly, anything you might be describing has many more details about it than you would ever want to write down or a reader would want or need. To verify this just look around. For instance, pick up a pencil and look it over carefully. How many factual descriptive details can you find? Color? Look carefully and you will find several different colors. The pencil's shape, length, sharpness, hardness? Get the idea? With a bit of thought you should be able to come up with fifteen or more distinct details.

Certainly there is no need to find, let alone list, fifteen or twenty different details about a lead pencil. A written description containing all these details would not only be boring to write and to read, but the unnecessary detail would probably get in the way of clarity. Your close observation of the pencil, however, has not been in vain. It should lead you to recognize that more complicated subjects than pencils possess even more details. Further, you can realize the importance of selectivity. Try working out a brief description of your pencil. Four or five important details should do nicely. Now pick a different group of five less "characteristic" details. Supplied with only this second group of details, even someone who uses a pencil regularly may not recognize that the object you are describing is a pencil! As you can see, then, an effective technical writer must select the most distinguishing details for descriptive writing.

Photographic Details

How do you decide which three or four details to use from a possible fifteen? Not just any three or four will do. Selecting the best three or four—or however many details you decide are necessary—can be done in either of two different ways: *photographically* or *impressionistically*. The photographic approach is the one you will likely use more often.

Photographic details are essentially what the name implies. They merely describe what is there—good, bad, or indifferent. Just as a camera can show what is there—masking, coloring, or exaggerating nothing—so can an effective photographic description.

To begin a photographic description you must observe your subject carefully. Study the subject just as you did the lead pencil in the example. To make sure you do not miss any potentially important details, jot down a list of all you observe. This list will help ensure that you do not forget any details, and later it will also help you to determine how best to organize them.

After studying your subject and your list of details, try to determine which details are most prominent. Which ones would a camera pick up? Which ones would help a reader form an accurate picture? How many of these characteristics you decide to use will depend upon the circumstances of your writing. Determine, as discussed in Chapter 1, who your reader is and what he or she wants the description for. Also, as common sense would indicate, larger or more complex subjects require more details than smaller, simpler ones. You will also find that some subjects have a few details that are so prominent and so unique that just those few will draw an effective word picture. Other subjects without such startlingly recognizable details obviously will require longer, more thorough descriptions.

Consider, for instance, the descriptive details you would need to draw an effective word picture of a file cabinet. An ordinary four-drawer gray or beige metal cabinet would require few descriptive details. A new, less common model, such as those which have files running crosswise and hanging from braces, would require many more.

The best way to check on the adequacy of your details is to show them to someone of the same general background as your prospective reader. If that person can identify the subject from your details, you will know you have done a good job. If not, you had better recheck your details and add or change a few.

Below are two lists. The first includes details about a particular ballpoint pen (the one used to write the first draft of this text). The second list includes only those details needed to write a good photographic description. Later in the chapter you will find an actual description of the pen. Notice that the first list of thirteen points actually includes over thirty details.

My Anson Pen
1. barrel mainly chrome plated
2. black ring about ⅔ way down
3. brushed chrome oval ½″ by ¼″ on top end of clip
4. otherwise ordinary shiny chrome clip
5. several areas of tiny lines running lengthwise
6. area of lines running around barrel just down from tip
7. ¾″ at tip only really shiny part of barrel
8. end opposite tip soft shiny blue plastic
9. five notches where chrome and blue meet
10. "Anson made in U.S.A." printed on underside near metal
11. lettering so tiny that one wouldn't notice it unless looking for it
12. no obvious buttons to retract point
13. ordinary size—5⅛″ × ¼″ diameter—slight taper on ends

My Anson Pen—Significant Details for Use in Description
1. standard size
2. generally chrome plated but not really shiny because of lines
3. black ring near center
4. shiny blue end
5. large brushed chrome oval on clip

With just these five details you should be able to quickly pick this particular type of ballpoint out of a large stack of pens.

Impressionistic Details

If using photographic details is like using a camera to photograph a subject candidly, exactly as it appears, selecting impressionistic details is like painting a picture to emphasize only those details that will create the impression you wish to convey about your subject. When you look for impressionistic details you select only those that fit the impression you want to give. In other words, if you and two or three classmates were to write photographic descriptions of the same subject, say a particular political candidate, your descriptions should be very much the same. They would probably include a matter-of-fact description of the candidate—height, weight, and other physical details, family, place of residence, background experience, and so on—a rather cut and dried, statistical sort of description. On the other hand, if you and your classmates were to write impressionistic descriptions, each description might be almost entirely different, depending upon what impression of the candidate each student wanted to con-

The key factor in producing a well-designed piece of technical communication is the writer's ability to exercise discrimination and good judgment when selecting details.

vey. You might do a "snow job," selecting only details to make the candidate seem wonderful, while a friend could do a "hatchet job," selecting only details to make the same candidate seem repulsive.

As a rule, most technical communication calls for photographic description, because it presents the most realistic, most nearly objective picture. In many situations, though, such as writing advertising copy, sales letters, or features in organizational newsletters, impressionistic descriptions are more appropriate and should be used. With either photographic or impressionistic descriptions, your discrimination and good judgment in choosing details are the key factors.

Not all impressionistic descriptions have to be absolutely positive or absolutely negative; actually very few are this extreme. Most of them are based on a specific impression that is not necessarily good or bad. For instance, the ball-point pen mentioned before could be described around each of the following impressions: sleek, clean, dull, chopped-up. Below are lists of details for each of these four descriptive categories:

Positive Details	**Negative Details**
sleek	*dull*
5⅛″ long, ¼″ diameter	mostly chrome
chrome plated	only tiny blue and black variation
small blue end	little of it shiny
vertical lines	
clean	*chopped-up*
chrome	shiny chrome, brushed chrome, lined chrome
vertical lines	long and thin
black and blue accents	black band, blue end
brush accent on clip	tapered ends

Selecting an impression is essentially a matter of personal taste. Ordinary honesty, of course, dictates that you should not do either the "hatchet job" or the "snow job" mentioned previously. A reader has the right to expect your impression to be sincere. You might think of the ballpoint pen in question as dull or chopped-up while a friend might think of it as sleek and clean—an honest difference of opinion. Your descriptions would be almost contradictory, but honestly so.

One final caution is necessary: When we think of slanted, biased descriptions we almost automatically think of heavily charged language, words that show the bias of the description. Although this is generally true, not all impressionistic descriptions sound that way. You can use plain, simple, unemotional language to make a description sound like a very straightforward, photographic one; yet if you have carefully selected your details impressionistically, the effect will be impressionistic. The careless reader will not even notice and will be deceived; the careful reader will realize being deceived and will resent it.

If you choose to describe impressionistically, do not attempt to disguise the fact by using language that sounds photographic (objective). Above all, be careful not to unknowingly mislead your reader by allowing careless selection of details to inadvertently give an impressionistic slant to a supposedly photographic description. Also, remember when you read that slanted details can give an impression even when the words used do not announce it.

Photographic Versus Impressionistic Language

Many people do not use the terms *photographic* and *impressionistic* when they speak of description; instead they talk of *objective* and *subjective*. *Objective* corresponds to factual, impersonal, statistical, and unopinionated; *subjective* suggests personal, opinionated. Objective, factual-sounding words are used in photo-graphic descriptions; subjective, judgmental-sounding words in impressionistic descriptions. Compare the pairs of terms that follow for a good idea of impressionistic (subjective) versus photographic (objective) details.

Impressionistic (Subjective)	Photographic (Objective)
tall	6'7''
fat	340 pounds
luxury	Cadillac
old, beaten-up jalopy	'53 Chevy with a fender missing
horrible catastrophe	$3,000,000 damage, 37 injured
mansion	$350,000 house
fuzz	police officer
defender of law and order	police officer
old woman	82-year-old woman
conservatively dressed	wearing a dark suit and tie
well-groomed	clean-shaven with combed hair

Here are some comparative sample sentences describing our ballpoint pen:

Photographic: It is of standard size—5⅛'' × ¼''.
Impressionistic: It is long and slender.
Photographic: It is chrome plated with a blue plastic tip.
Impressionistic: It is plated in sleek, shiny chrome with a decorative deep blue tip.
Photographic: No retractor button or mechanism shows.
Impressionistic: The clever design eliminates any visible retractor mechanism.

Organizing the Description

Descriptions are one of the easiest forms of writing to organize. Often you can merely list details naturally as you select them and they will almost organize themselves. But it is not always that easy.

Photographic descriptions are generally organized in some sort of spatial or physical arrangement. You might go from left to right or from top to bottom, inside to out, front to back, or even in a circle. Often the subject will be so shaped that a particular pattern is almost forced on you. If so, use it. You might have a choice of several patterns, each of which would work. Give the decision some thought and choose the one that seems best to show the total structure, each detail fitting together to make a whole. You can only go wrong if you fail to use a pattern.

Sometimes, even in a photographic description, you will not want to use a purely spatial pattern. For instance, a spatial-order description of a ballpoint pen—going from one end of the pen to the other—would not really be needed if your reader has seen ballpoint pens before. Rather, to give the reader a good general image, you might begin by mentioning that the pen is a normal-sized and normal-shaped ballpoint. Then list the details that distinguish this particular pen from other pens. This pattern, the "familiar-unfamiliar" pattern, is especially useful for describing a familiar machine or tool that has new or unusual features. It is also effective in describing a seemingly unfamiliar object that has prominent similarities to an object familiar to the reader. For example, assume you are preparing a description of a computer terminal for an employee who is transferring from the secretarial to the computer staff. An effective beginning for your description might be: "If you look for an IBM Selectric typewriter when you enter the data center, you will spot the NCR 260 terminal immediately." Then, of course, you can proceed to compare the typewriter and the terminal.

When describing large, complex subjects you might use the "whole-parts" pattern. Begin by mentioning the general size and shape of the whole, then describe each part or section, using an appropriate pattern for each part. That way the reader can get a good overview of the whole subject and can readily fit each piece into that whole.

Whichever pattern you choose for your photographic description, keep in mind that some clear, understandable pattern is necessary. No description of individual parts or details, no matter how carefully prepared, can work unless the reader can see how the parts or details fit together.

Impressionistic descriptions can be developed with these same patterns. Remember, however, that you should have a definite impression in mind and that you should select both language and details especially to create that impression in your reader's mind. From the beginning try to establish clearly the impression you intend, yet get right into the description. Do not waste a sentence and risk boring your reader right off by stating your impression flatly. Communicate your intended impression by careful, thoughtful presentation of selected details. Once the impression is clearly established, proceed with the description using one of the patterns: spatial, whole-to-parts, or familiar-unfamiliar.

Following are two examples of openings for an impressionistic description of a snack bar area of a student center. Notice that two quite different impressions can be developed about such an area—"warm and alive" and "loud and chaotic."

> After spending an hour or two in one of Mudville's cold, sterile classrooms, walking into the student snack bar is like being reborn. You hear people talking, laughing, and even singing. Here are warm, live human beings, not the. . . .
>
> A prospective freshman wandering into our student snack bar for the first time might get some second thoughts about attending Mudville. Here are wall-to-wall bodies, singing, yelling, cussing, boogying, making a shambles out of the place. . . .

Now here are three descriptions, the first photographic, the second and third impressionistic, of our ballpoint pen. Notice that the photographic description begins by describing the overall size and shape, then focuses on the smaller details. Each impressionistic description begins with a clear, definite statement of the dominant impression.

My Anson Pen

My new Anson ballpoint pen is of a normal size (5⅛" long × ¼" in diameter) and shape. It is chrome plated, with tiny grooved lines running lengthwise down the barrel and horizontally around the tip. There is a ⅛" black band around the center of the barrel and a blue plastic inset in one end. Its only other prominent feature is a ½" × ¼" brushed chrome oval plate on the clip.

A Good, Clean Pen

My new Anson ballpoint pen has the same kind of clean, strong lines one might associate with an automobile such as a Mercedes-Benz. No gingerbread to distract the eye; it is covered in good, solid chrome plate. There is a tiny black band around the center and, for accent, a shiny, blue plastic plug in one end. Final touches of elegance are the brushed chrome oval plate on the clip and tiny grooved lines on the barrel. This is definitely a clean, no-nonsense pen.

A Chopped-up Mess of a Pen

Unlike most so-called "good" ballpoint pens, my new Anson is not at all attractive. Rather it looks as though several designers could not agree on one design, so they each threw in a feature or two. To begin with, it is just a bit narrower than competitive pens, but this quality is then exaggerated by vertical grooves running most of the way down the barrel. And just as the eye gets used to these, it is jarred to see a one-inch band of horizontal grooves near the tip. Making the design even more awkward are the prominent black band around the middle and a gaudy blue plug in one end. But even these gauche decorations look good compared to the ridiculous large oval plate protruding from the clip. The mixing of three clashing chrome finishes completely destroys the pen's appearance. Most of the barrel has a soft chrome finish highlighted by the grooves, but one end is highly polished shiny chrome while the oval on the clip is brushed chrome. No one feature is that bad by itself, but the features do not fit together.

Narration

Unquestionably the narrative is the most common writing pattern used both in and outside of technical communication. Novels, movie and television scripts, personal anecdotes—all are narratives: they tell what happened. The narrative pattern in technical communication fulfills the same function. Log books, weekly reports, progress reports, minutes of meetings, travel vouchers, many letters and memorandums—even the formal report—all tell what happened, all are narratives. You will find the following narrative patterns helpful in countless job-related writing situations.

Basic Narrative Patterns

The basic narrative pattern is exactly what you probably expect—chronological order, beginning with the first event to be covered and going straight through to the most recent event. If you are reporting on a month's work, you begin with the first working day and end with the last. If you are reporting on a business meeting, begin with the meeting being called to order and work straight through to the adjournment. The straight chronological pattern is probably the easiest to write and to read. Use it frequently.

The parallel chronological pattern, which relates several simultaneous chains of events, is a frequently used variation of the straight chronological pattern. For instance, if you were a firefighter, you might be expected to report everything of significance you had done each month. If your duties were varied, with parts of each day spent on each of several activities, or if you happened to be working on some sort of major inspection or fire-safety project while keeping up with your regular duties fighting fires, a single chronological narrative might be confusing to the reader, who would have to switch back and forth mentally as your report recorded first one of your activities and then the other. Two (or more) parallel chronological patterns—one narrative for each activity—would be far easier to follow and understand. Your reader could still determine your total monthly accomplishments by combining your several narratives.

The parallel chronological pattern is also excellent when a supervisor must narrate the work accomplished by a department whose members engage in different activities. For instance, a construction superintendent might have to report on work done by eight or ten different crews, each working on its own specialty. Parallel narratives, one for each specialty, might be called for. Or say you were a department head of a large laboratory. If several different types of analyses were done regularly, it might be a good idea to narrate each type separately.

Another variation from straight chronological order is the flashback pattern, best known for its use in fiction. In a book or movie, for instance, the central figure may be seen first as a young adult and finally as an elderly man, but seen in between as an adolescent or even an infant. Such flashbacks keep the emphasis on one time period but allow earlier events to be included. This pattern is not as common in technical writing as it is in fiction writing, but it is sometimes quite effective. In a situation where you want to emphasize recent events but want to mention much earlier events briefly, use the flashback pattern. Begin the narrative with the first in the continuous series of recent events. Then find a spot where you can flash back smoothly without awkwardly interrupting an event or a closely related chain of events. At that point briefly relate the earlier events before continuing the basic narrative. For instance, in writing a report on planning a building project, you might flash back to an unsuccessful attempt made ten years earlier to begin the project.

When using the flashback, keep two cautions in mind: First, try to find a natural break in which to insert your flashback; avoid interrupting the first sequence of events or any especially important sequence. Second, use the flashback pattern only when your flashback is relatively short compared with the basic narrative. If the flashback events are almost as long or complex as the basic events, place them in the order of their occurrence and use the conventional chronological approach.

If you can become reasonably competent at writing each of these three narrative patterns you should be able to handle nearly any narrative you are likely to face. Choosing among the three should not be too difficult. Most of the time straight chronology will work satisfactorily. At times you should realize that too many activities are involved for one narrative and you can easily switch to a parallel series of chronologies. Once in a very great while you will find a narrative which seems to be disjointed; some events will have occurred long before most of the others. Just switch to the flashback pattern.

Straight Chronology

Eh 157-Seq. 1080, January 14, 1980
(Report of a typical class meeting)

When the period began, I called the roll. All were present. I then checked for papers to be turned in and collected four revised abstracts and one rough draft of a formal report. The class then began discussing resumes. Since we had previously discussed the philosophy behind resumes and their functions, we started in on the mechanics of constructing one. "Personal Data" and "Educational Background" were covered quickly with no major problems. But "Work Experience" took most of the period. Several students had provocative questions about recommendations to former employees. Debate centered on the practice of some companies of giving employees being fired good recommendations and refusing to give recommendations for employees who quit. The period closed with my giving the class an assignment to begin rough drafts of their own resumes.

Parallel Series

Professional Activities—Weekly Report, July 8–12

During the week I worked on three specific activities: teaching two technical writing classes, writing my new textbook, and discussing some faculty forum activities. I spent four and one-half hours in class on Monday and marked two sets of descriptive and definition papers. I also spent one hour reviewing abstracts. On Tuesday I lectured on abstracts to both classes. On Wednesday I helped students write abstracts and introduced them to technical letter writing. I also marked another set of papers. On Thursday and Friday the classes proceeded in the discussion of technical letters, and I marked two more sets of papers.

On Monday afternoon I roughed out my outline for Chapter 2 of my proposed text. Tuesday evening and Wednesday afternoon were spent writing a section. On Thursday evening I revised this section and put it aside for typing.

Faculty forum activities took several hours of my time too. On Monday several of us discussed faculty involvement in new campus development and faculty administration relations. The forum met on Wednesday to debate these subjects. Nothing was decided. Informal discussions followed on Thursday and Friday.

Flashback

Progress in Writing New Textbook

Serious work on the text began on May 5 this year. The first item prepared was a detailed outline. Three revisions and several consultations with colleagues later, I was ready to begin writing. This was the culmination of two years of planning and weighing alternatives. In late 1977 I began sensing the need for a more comprehensive text covering both technical writing and technical English, but I could not decide exactly how to combine the two. Finally, in February of this year I discussed the idea with several publishers, and a plan for the book began to emerge in my mind. Two more months of hard thinking and I was ready to begin. In mid-May I began developing Chapter 1. Completion took several weeks. Then I spent the month of June on Chapter 8. Several days were wasted in early July as I tried unsuccessfully to get Chapter 2 going. I finally got on the right track on July 8, and Chapter 2 is now coming along well.

Selecting Narrative Details

As you might suspect, selecting details is just as important in composing a narrative as in composing a description. Any event or series of events you might narrate has many more details than you can possibly use. Your choice of details can show the event favorably, unfavorably, or somewhere between. Consider for a moment the most recent party you have attended. With the possible exception of a few parties—which not even Shakespeare himself could make sound exciting—you could probably make the party sound like the best one in history merely by selecting favorable, exciting details. On the other hand, you could easily make the same party seem dullness itself by selecting other details.

The vast majority of technical narrations are photographic (or objective). They record the important facts, those that show an event as it really was. This can never be done perfectly, since total objectivity is impossible when someone has to record the details and determine which ones to use. But perfection can be approached, and you should work toward that end. Begin by listing as many factual details as you can, and avoid judgments. Double-check your list to make sure that you have not left out anything important. Then go through the list and select the most important details using such criteria as "How long?" "How many?" "How much?" In other words, select those details that tell what lasted longest, involved the most people or the most money, or caused the most change. Rely on your common sense; judge as you believe the average intelligent person would. You can never be totally objective, but you can be fair.

Occasionally you might wish to narrate subjectively, impressionistically. If so, select details as you would in developing an impressionistic description. First determine your impression; then carefully select only those details that help convey it.

Below are three lists of narrative details about the first ten minutes of a meeting. The first includes all details observed; the second is an objective list; the third, subjective.

Complete

1. Called to order by B. Osborne
2. Secretary's report by M. Parker
3. Five minutes' discussion

4. Motion by Paulson, second Burgett
5. Approved by voice vote
6. Noticeable nays
7. Report of Activities Committee by M. Dahlberg
8. Group picnic scheduled for Aug. 29, 6–9:30
9. Families invited
10. Many games
11. Kentucky vs. Maryland fried chicken
12. Chicken brand debated from floor
13. Osborne ruled such debate out of order
14. Report accepted by voice vote with loud nays.

Objective details
1. Called to order
2. Secretary's report
3. Discussed and approved
4. Report of Activities Committee
5. Picnic for Aug. 29 outlined
6. Brief discussion
7. Report accepted.

Impressionistic details
1. Some discussion of secretary's report
2. Vote not unanimous
3. Report by Activities Committee
4. Chicken brand debated
5. Ruled out of order
6. Many nays when accepted.

The Language of a Narrative

Just as carefully selected details can make a narrative either impressionistic or objective, so too can the choice of language that is used in the narrative. As in an impressionistic description, emotional, judgmental words make a narrative impressionistic or subjective; and straightforward, factual words make it photographic or objective. Take a close look at the following lists of terms and note the contrasting feelings (or lack of feelings) each list evokes.

Subjective, emotional	versus	Objective, factual
debated		discussed
admitted		said
avoided		made no comment on
ambled		walked
grilled		questioned
gobbled		ate
dude		man
lady		woman

chick	young woman
teenybopper	teenage girl
liberated	released
plastered	intoxicated
eyed	looked at
clobbered	beat
cliffhanger	3–2 game

Chapter 6 offers more suggestions for selecting the right word. Two sample narratives—one objective, the other subjective—follow, each describing the events at the same meeting. Notice that they sound almost nothing alike.

The July 10 Meeting

The meeting was called to order and the secretary's report read. Her report was discussed for some five minutes, then approved. Next came a report from the Activities Committee. The major item mentioned was the annual picnic scheduled for August 29. A brief discussion followed before the report was accepted.

The July 10 Meeting

The first significant event of the meeting was a five-minute argument about whether to approve the secretary's report. When it was finally approved, many observers felt that the chorus of "nays" was louder than the "ayes," but it was approved anyway. This episode was nothing compared to the next. Tempers flared as members fought over whether to follow the Activities Committee's recommendation to have Kentucky Fried Chicken at the August 29 annual picnic or to substitute Maryland Fried Chicken. The club president finally had to rule the debate out of order. The ensuing vote in favor of the committee's report was again almost too close to call.

Changing a few details and using strongly emotional language can make a rather normal meeting sound like a street-gang confrontation!

Definition

This section will discuss working definitions, which explain to the reader how a term is being used in a particular context. It will also look at general definitions, which give the reader a good, clear understanding of the term in all contexts.

Working Definitions

A working definition is designed for use only in one context—the particular report, memorandum, or paper in which it appears. Conceivably, a single term could be defined in almost entirely different ways in two different reports. This is unfortunate; but if each definition is clear in the work in which it appears, communication will still take place. A working definition does not pretend to be thorough, nor for that matter even objective. What it does do is show readers what is meant in a specified situation. It says to the reader: "Every time you see term *x* in this paper, here is what you should take it to mean."

Several types of words require working definitions. Probably most common is the highly technical term that would simply be unfamiliar to the reader. Usually it is a good idea to avoid using such terms by substituting more familiar equivalents or by rewriting the sentence in a different way. Sometimes this cannot be done, and you are forced to use the technical term. Then you have a choice: Write a short translation that gives little real insight into the concept (a working definition), or write a longer explanation that gives the reader a good knowledge of the concept behind the term (a general definition). If the reader has no immediate need to learn and apply the entire new technical concept, your best choice is a working definition.

A second situation that calls for incuding a working definition where a familiar term is used in an unfamiliar way. Such terms are often called "shoptalk" or "jargon." For example, the word *hog* in jargon may refer to a large Harley-Davidson motorcycle. A writer using *hog* with that meaning would be obliged to inform his readers of the irregular word usage by including a working definition. Similar to this are the many abbreviations, acronyms (initials that spell out a word and are usually pronounced as a word, such as CORE or NOW), numbers, or initials used instead of the equivalent word or words. Medical fields and government agencies, for instance, make heavy use of initials, acronyms, and abbreviations.

One particular class of words is especially troublesome: the abstract words, like *love, hate, freedom,* and *justice,* whose meanings philosophers have been arguing about for centuries. The meanings of these words are hard to pinpoint, and there may be several almost equally common meanings, any one of which could make sense in a given situation. Your best course is to avoid using such words entirely. If you must use them, then define them with a working definition.

Closely related are words such as *communist* and *democracy,* which have precise meanings to economists and political scientists but a variety of meanings to the rest of us. Then there are words that have several different, almost contradictory meanings. One such word is *hippie.* To some persons, *hippie* designates any young man with shoulder-length hair. To others, a man with certain philosophical notions having nothing to do with hair might be considered a hippie. With two such different understandings of the word, no effective communication can take place. Avoid such words or translate them with working definitions.

A working definition need not follow a particular form. Frequently, one or two words will do. Other times you may need a phrase, a clause, or even a sentence or two. Working definitions are seldom difficult to write and often can be written after only a few seconds' thought. The trick is to remember to use them when they are needed.

General Definitions

The major fault with most technical definitions is that they contain no pattern at all; rather, the writer has just listed characteristics. This often works, as the reader may eventually get a notion of the term's meaning. But sometimes little or no meaning at all comes through.

One good, workable pattern for a general definition consists of the following parts:

1. The term is placed into a class.
2. Characteristics should differentiate the term from others in the class.
3. Synonyms, drawings, and other aids are furnished.

Placing the term to be defined into a class is not as simple as you may at first think. If you are able to define the term, you can usually come up with a class very quickly. Often, however, the first class that comes to mind is not the best. The purpose in placing a term in a class is twofold: to give the reader a quick general notion, a starting place, and to cut down on the amount of detail it is necessary to present. If the class is too broad, the starting point you give the reader is not the best you could supply, and you will need too many details to make your meaning clear. Conversely, if the class you choose is too narrow, you will not have to present many details, but the reader will not understand your general definition. Obviously, a wise choice of class size and scope is a key to writing a good definition.

Suppose you had to define a Datsun 280-Z (really define, not explain). You could choose from among the following classes: thing, vehicle, car, Japanese-made car, foreign sportscar, updated 240-Z. Any of these classes could be made to work, but some would work better for some readers than for others. If your reader is an American familiar with the 240-Z but unaware that Datsun has altered it slightly and changed its name first to 260-Z and then to 280-Z, you could start by calling it an updated 240-Z. If the reader is an American who has seen many 240, 260, and 280-Zs as well as other foreign sportscars but has paid little attention to any of them, you could begin by calling the 280-Z a foreign sportscar. If, in the extreme, your reader has recently returned from years in the Antarctic, you had better call the 280-Z an automobile. The principle illustrated is this: Use the smallest class that you can be sure is familiar to your reader.

Once you have established your term within a class, begin differentiating it from other members of the class. If, for instance, you had classified your 280-Z as an updated 240-Z, you would merely mention the changes made. You would point out that the primary change was to increase the engine displacement from 2,400 to 2,800 cubic centimeters. Then you might mention that this resulted in little increase in power because of the increase in emission controls. Then you could mention the minor changes in the suspension system and in the interior decor. And finally you should mention that by and large the 280-Z is basically the same car as the 240-Z.

If you had classified the 280-Z as a foreign sportscar, your distinguishing details would be more basic. The 280-Z is made by Nissan motors, a Japanese firm, under the name of Datsun. You could call it a moderately priced sportscar, selling for roughly $10,000—more than such popular small European sportscars as MG, Triumph, and Fiat X19 but much less than such European imports as Porsche, Ferrari, and Maserati.

If you had labeled the 280-Z an automobile, you would have to explain the concept of sportscar before you could even begin to focus on the 280-Z. You would have to point out, among other features, the two seats, low-slung design, and high-performance handling.

After giving enough details to distinguish your term clearly from other members of the class, supply whatever additional details you believe are necessary to give the reader a good, usable understanding. Definitions will vary considerably in length. Terms denoting highly complex processes will usually need rather long definitions. Technical concepts, theories, and principles can be explained in relatively few details if you stick to a single concept or theory. However, if you choose to explain how a term fits into a broader subject, you can write almost endlessly. For instance, a computer programmer could write a very effective definition of a given computer language, such as "Cobol," in one good paragraph; going beyond this to explain the broader concept of computer languages in general might require several pages.

The only way to determine just how much to say is to carefully answer the preliminary questions "for whom?" and "for what purpose?" If you can identify your reader and can determine just why he or she needs a given definition, you can usually determine just how much information to provide. When in doubt, remember these two tips: too much detail is better than not enough, but details that do not clarify anything can get in the way.

Do not overlook any type of detail in your effort to make your definition clear. You may want to use the term in a sentence or two, just as most dictionaries do. You may find synonyms helpful. For instance, in defining the term *Allen wrench* you might find the synonyms *hex wrench* and *hex key* helpful because many readers may be familiar with the Allen wrench but know it by these other names. A sketch may be useful. Some people, for instance, would recognize an Allen wrench by sight but would not know it by name. Even if the reader does not recognize the tool, the sketch can be helpful in another way: If what you are defining is very complex or very unusual, a sketch gives you a reference point for your explanation of various parts or features.

Sometimes you will find a list of types or subclasses useful. To introduce one, make a statement such as "Some of the more common types of *x* are *a, b, c,* and *d.*" This is especially useful when you are defining a broad term such as a family of trees, a classification of crimes, or a class of fires. Your reader may well be familiar with a subgroup, and that familiarity will help him to understand the broader term more easily.

In brief, once you have established your basic definition, placed your term in a class, and differentiated it from other members of the class, look for any additional helpful information. Use as much as you need to help the reader understand your term.

The three definitions that follow show varying techniques required by the differences in the terms being defined.

> *Avant-garde* is used in interior decorating as it is in fields such as literature, music, art, and fashion to designate an entirely new approach rather than anything more specific. A simple definition might be "experimental, trying new ideas and techniques, and ignoring established traditions." Most of today's established and accepted decorating styles were once considered avant-garde, really extreme or daring. A good example is the use of cushions on platforms or on the floor as an alternative to conventional seating. As recently as twenty years ago this was considered extreme or daring. Today, of course, it is well accepted.

> The *median* is a commonly used method of describing a distribution of numerical statistics. It is specifically the figure at the midpoint of a distribution of statistics above which and below which there are the same number of cases. Consider, for instance, a distribution of the following scores: 72, 74, 87, 91, 93, 96, 99. The median is 91, with three scores higher and three scores lower. In the distribution 11, 11, 12, 14, 15, 16, 17, 18, 19, 20, the median is 16. With an even number of scores, the median can be determined by finding the value halfway between the two middle scores. For instance, the midpoint of scores 6, 7, 8, 10, 11, 11 lies halfway between the two scores of 8 and 10. The midpoint, therefore, is 9. No matter how large or complex the distribution of statistics, the principle for determining a median is always the same: determine the exact middle score in the distribution.

> The pigmy rattler is the smallest of the rattlesnakes, measuring from twelve to twenty-four inches. It is quicker tempered than other rattlers and often coils and strikes repeatedly. Its bite is painful and should be treated as any other poisonous snakebite, but no deaths have been known to result from it.
>
> The pigmy is native to nearly all of the southern United States and is most commonly found in wet marshy areas or near large lakes. It is usually dark gray with large black blotches on its back and sides and a rust-orange stripe running down its back. It has the wide, pointed head with vertical pupils and facial pits common to rattlesnakes. Its tail is long and usually tapers to very small rattles.
>
> The pigmy rattlesnake is also known as the ground rattlesnake; its scientific name is *Sistrurus Miliarius.*

Comparison and Contrast

Technical people are often called upon to compare and contrast two or more machines, products, or ways of doing some operation. Two comparison-contrast patterns—the "one-other" and the "point-point" patterns—are very helpful in such situations. Suppose you had recently tried a new piece of equipment for a few days and had to recommend to your boss whether to replace all your old machines with these new ones. Using the one-other approach, you would discuss the new machine, then you would discuss the old (or vice versa). Using the point-point approach, you would discuss the old and new together, aspect by aspect. The brief models below illustrate the two approaches. First are brief lists of characteristics of two kinds of grass, then a point-point comparison, followed by a one-other comparison.

Point-Point Comparison List

St. Augustine	Bahia
broad leaf, thick	narrow leaf, open
dark green	medium to light green
sod or runners	seed or sod
worms and chinch bugs	few pests
shade resistant	sun or partial shade
heavy fertilization and water	medium fertilization and water

Point-Point Comparison Paragraph

St. Augustine Versus Bahia Grasses

As potential lawns for our new building site, both St. Augustine and Bahia grasses have distinct advantages. St. Augustine is broad leafed and thick, whereas Bahia is narrow leafed and relatively open. St. Augustine is a rich dark green, Bahia a much lighter green. St. Augustine will withstand shade much better than Bahia, which requires at least partial sun. Bahia has much greater resistance to pests, with St. Augustine being especially susceptible to sodwebworm and chinch bug damage. Bahia requires less water and fertilizer and is much cheaper to establish since it grows from seed, whereas St. Augustine requires sod or runners.

One-Other Comparison Paragraph

Bahia Versus St. Augustine Grasses

Either St. Augustine or Bahia grass would make excellent lawns for our new facility, with each having definite strengths and weaknesses. Bahia is inexpensive, as it grows from seed. It is also extremely resistant to pests, and it requires relatively little watering or fertilization. The main disadvantages of Bahia are its need for at least partial sun and the mediocre appearance caused by its being narrow leafed and light colored.

St. Augustine, on the other hand, is thick growing and broad leafed. It is a rich dark green, which enhances its appearance. Furthermore, it will grow well in open sun, partial shade, or even nearly complete shade. On the negative side are its requirement for much water and fertilizer and its susceptibility to damage from sodwebworms and chinch bugs. It is also expensive since it must be grown from sod or from runners.

Use the one-other approach when you want to compare total impressions rather than specific characteristics. When you want to compare or contrast specifics, use the point-point approach or use a table. Tables can and should be used for comparisons. In fact, any time you are considering a point-point comparison, also consider using a table.

Partition and Classification

Many explanations are so long or complex that they must be broken down into a series of smaller explanations. When presented part-by-part or phase-by-phase, long, complicated processes are easier to write and to read. Another common explanation requiring this kind of partitioning is the explanation of a structure. Suppose, for instance, that you were explaining to prospective employees your company's insurance plan. Chances are the plan would be complicated, composed of hospitalization insurance, doctor's insurance, major medical insurance, and perhaps even dental insurance or income-protection insurance. To make your presentation comprehensible, you should partition the subject—our insurance plan—into subgroups: hospitalization, major medical, and so on.

Classification achieves the same result as partition, but it works almost exactly opposite. To partition, you divide one complicated subject into several less complicated subjects; to classify, you group a large number of individual items into a smaller number of groups. For instance, in preparing a talk to a group about the job opportunities in your profession, you might discover that as many as fifteen or twenty different types of positions were available. To keep that many positions straight, you could classify them into three or four subgroups. Then you could explain each group more easily to your listeners.

You will find classification and partition used throughout this text. Good examples of partition are the explanations of the resume in Chapter 14 and the formal report in Chapter 10. Each explains a complicated document by breaking it down and explaining each part. Chapter 8 furnishes a good example of classification. Letters come in more different types than one can count; yet they have been explained through classification of many types into a few basic groups. Some suggestions for classifying or partitioning follow:

1.   Classify or partition according to some logical basis. Most groups can be classified according to many bases. Find one that seems to work and apply it. For example, employees in an organization can be classified in many different ways—hourly versus salaried, permanent versus temporary, full-time or part-time, and so on. No one basis is automatically the best in a given situation. Find the one that works in your situation.

2.   Be sure that your basis covers all parts or items, and apply it consistently. Do not mix bases. For example, you could classify automobiles on many bases, but you should not mix them into such combinations as Fords, V-8s, mid-size models, and foreign-made cars.

3.   Once you have determined your set of parts or groups, look for a logical sequence in which to present them. For instance, if you were explaining aspects of your group insurance plan you might present your subtopics in most-common-to-least-common order, starting with hospitalization and ending with income-protection insurance. Or you might go from least expensive through most expensive. Just find a logical order and use it.

• EXERCISES

1. Select a place you are familiar with—a building, room, yard, or any other type of space—and jot down all the descriptive details you can about it. Now decide upon an impression you want to create and select only the details from your list that will help you develop that impression. Make a second list of details, this time selecting only those that would work best for a photographic presentation. Then write each description, selecting objective language for the photographic description and subjective language for the impressionistic presentation.

2. Rewrite each of the following in objective language:
   a. Her hair always looks beautiful.
   b. She works too hard.
   c. He is a very easy teacher.
   d. Cats are easy to care for.
   e. His car is very economical.
   f. The room was a mess.
   g. Toddlers are so inquisitive.
   h. The beach was overcrowded last weekend.
   i. This room is way too hot.
   j. The receptionist was rude.

3. Find subjective equivalents for each of the following:
   a. instructor
   b. television set
   c. replied
   d. supervised
   e. cigarette
   f. lawyer
   g. doctor
   h. cluttered
   i. discuss heatedly
   j. suggest firmly

4. Recall some recent event you participated in or observed, something out of the ordinary, so that you can remember many details. List all significant details about the event as accurately as possible. Then select those you would need to narrate it objectively (photographically). Now select an impression of the same event and pick the details needed to narrate it impressionistically. Write the narrations, using the appropriate language for each.

5. Construct three working definitions. One term should be a highly technical one from some specialized field; one should be a slang, shoptalk, or jargon expression; the third should be an abstract word of the *love, hate, war, justice, truth* variety.

6. Now write a general definition. Select a highly technical term and define it fully and clearly enough for a nontechnical reader.

7. Select some common object around your room, home, or office; then choose a similar object of a different brand or model. Tapedecks, television sets, vacuum cleaners, any such common objects will work. Prepare lists of comparative

and contrasting details similar to those on page 50. Decide whether a one-other approach or a point-point approach would work better. Compose the comparison-contrast.

8. Think of an organization you are familiar with, such as a department store, fire department, large high school, or small business. Divide the organization into branches or divisions and explain how it is organized.

9. List fifteen bases of classification for the students at your school, not counting the very obvious ones such as year in school. Pick one basis and write a short paper explaining how students can be classified by it.

10. Find at least five bases of classification for each of the following:

    a. television sets
    b. singers
    c. computers
    d. Chevrolets
    e. pets
    f. instructors
    g. restaurants
    h. movies
    i. toothpastes
    j. hobbies

11. Partition each of the following into at least four basic parts:

    a. a typical schoolday
    b. a computer
    c. a wristwatch
    d. this textbook
    e. a bicycle
    f. a stereo system
    g. a Big Mac, Whopper, or the like
    h. a pencil
    i. registration at your college
    j. a tennis shoe

4

Process and Instructions

Process

    The Basic Pattern

Instructions

Exercises

Overview   *This chapter will examine two modes of communication especially important to technical people. A* process analysis *shows the reader how something works;* instructions *show how to operate it.*

Since most technical people work with or around electrical or mechanical equipment, explaining how that equipment works and showing how to use it are two forms of communication they must frequently use. Because of the importance of these two modes, this chapter will examine them closely.

Process

Process—also termed *process analysis, process description,* and *operational analysis*—explains how something works. It is one of the most commonly used modes in technical communication.

Before we can concentrate on telling how something works we must first make an important distinction: telling how a process takes place is not the same as telling how to do that process, although there is some overlap between the two. To make certain of the distinction, consider the operation of a manual-transmission automobile. Both types of papers could be written about this operation, but they would differ considerably. You could explain how to operate such a car—how to depress the clutch, put the gear shift in a certain position, and do the other steps necessary to start the car, get it moving, and shift gears. This paper would not give the reader any understanding of what happened when the clutch was pushed, it would merely point out that if the clutch wasn't depressed, the gear shift could not be moved. In short, the reader would know how but not necessarily why. On the other hand, a paper telling how the process works would explain how the clutch plates and cluster gears work. It would not necessarily tell the reader what steps to take to make them work.

You can see how a good set of instructions would imply much about the process and a good process explanation would imply even more about "how to do it." But neither would effectively substitute for the other.

The Basic Pattern

Papers telling "how it works" are written for many different types of readers. It might be necessary, for instance, to write such a paper for a customer, a community group, or some other layman who has little specific knowledge of the particular subject. You might have to explain a process to a new employee or trainee who has some knowledge of the field but is not a trained specialist. Or you may have to explain something to a colleague or supervisor who is expert in your field but who has never seen this particular process. So, when doing your preliminary study (as discussed in Chapter 1), it is especially important to determine as much as possible about your reader's knowledge of the subject and his or her purpose in seeking the information. Make your paper fit your reader.

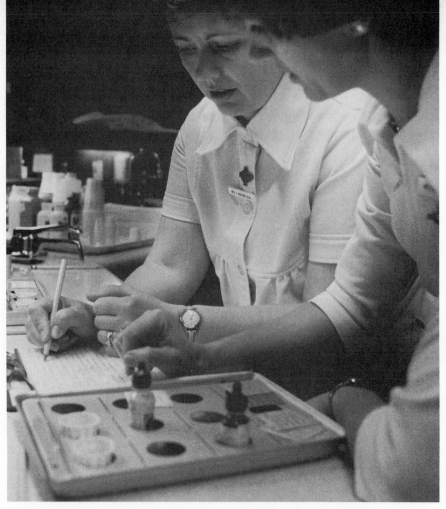

Employees frequently have to explain a process to someone new on the job. The *process analysis* is one of the most commonly used modes in technical communication.

A good explanation of how something works may have the following basic sections:

1. Definition and overview of the process
2. Working definitions
3. Descriptions of apparatus
4. Explanation of theory
5. Actual description of the process.

Most explanations do not need all five sections. Often some sections are omitted and others combined. This section will examine all parts so that you can use them when the need arises.

Begin with a simple, direct statement—a definition—of exactly what the process is. Obviously you cannot explain the entire operation in one sentence, but you can give the reader a good general notion of what kind of operation you are explaining. You might, for instance, begin with a statement such as "A voltameter is a machine used to measure electrical current," or you might say "A sphygmomanometer is the device commonly used to measure blood pressure." A beginning sentence such as these lets the reader know what the process is used for; it also identifies the field in which it is used.

One or two sentences will usually serve as an adequate introductory definition for a short explanation of a fairly simple process, particularly if your reader is a layperson. Longer, more complex processes require more thorough definitions, especially if the reader needs a complete, highly technical understanding of the process. Chapter 3 will help you write such introductory definitions.

Follow your definition with a two- or three-sentence overview or summary of the main phases or steps in the process. Unless you are absolutely certain that your reader is familiar with the general subject, use plain, nontechnical language for this section. Try to give a basic notion of what to expect—of what the more specific, more technical discussion to follow will be about. As with the definition step, if the process involved is long and complex, this section will be longer and more complex. But do not overdo it. Remember, your purpose is to introduce the process, to give a basic understanding. Later your explanation can get more technical and complete.

After you have written your brief definition and overview, you may want to define a few terms. You will probably find it best to identify them at the place where you first use them in your detailed explanation. Jot down any terms you can quickly think of and be alert for more as you go ahead with the detailed explanation. Later you can write out the working definitions.

You must decide how many terms need to be defined. If you can explain the entire process clearly without using any unfamiliar terms, by all means do so. But if you must use unfamiliar terms, use working definitions.

Describing apparatus in a process explanation is much the same as writing working definitions. Generally you will not need a section describing apparatus. If the process uses complex, unfamiliar apparatus, give the reader just enough of a picture of it to explain its use in the process. For instance, if you were writing about the common process for distilling water, you would need to describe the basic parts of the apparatus, but you certainly would not need to describe every rubber stopper or piece of tubing used. In fact, if you did try to describe the apparatus down to the last detail, you would probably confuse and lose the reader. Confine your description to just the basic parts that you will be referring to in your explanation.

Do not hesitate to use a simple sketch if you think it will help the reader visualize the basic parts and their relation to one another. Often a sketch with a few words identifying the different parts shown will explain the apparatus better than several pages of words. Whatever approach you use, remember that if the process you are explaining involves unfamiliar apparatus, your reader must understand it to understand your explanation.

An explanation of technical theory is the fourth preliminary part you might need in a process explanation. If your reader is totally unfamiliar with the subject, you will want to give some background theory of the technical or scientific principles involved behind the process. Use the same guidelines you do for working definitions or description of apparatus: give only enough information for the reader to comprehend the process. You are not trying to teach theory. That can be done elsewhere. A typical explanation might go like this:

> This process is based on two elementary scientific principles. First is the principle of convection, which basically means that hot air rises and cold air drops. Second is the principle of heat absorption; that is, the amount of heat waves absorbed by materials is influenced by the color of the materials. Hence, black roofs absorb more heat than white roofs and help keep a house warmer in winter and make it hotter in summer.

Do not explain theories the reader already understands; just mention them by name. And do not explain theories he or she does not need to understand. Unnecessary details may confuse and irritate the reader.

The heart of any process explanation, whether it is a brief, oral explanation to a co-worker or the body of a lengthy, written report, is the actual, step-by-step process or operation. Most processes can be explained best by a chronological approach, beginning with the first step involved in starting the operation and ending with the final step completing it. Check your steps carefully to make certain that none are omitted and that you get them in the proper order.

Like the events in a narrative, the steps in your process that happen first are mentioned first—but use no flashbacks here! The steps in a process must be in sequential order, one step leading to the next, because the process would not work if steps were out of order.

Some processes do not have one starting point or one obvious final stage. Instead they function as cycles, with one step leading to the next, around and around. Some such cyclical processes (such as the well known Krebs cycle, or nitrogen cycle) are never ending. Once set in motion they seem never to stop. (In fact, if the Krebs cycle stopped, plants, animals, and eventually humans would starve to death.) To explain any such cycle, select a starting point at what seems to be one extreme and work your way through the cycle to that point. Generally there are several potential starting places, and any one will do. Some cyclical processes do have a definite, logical starting place. Take the functioning of typical four-cycle automobile engines (rotary engines are a different matter), for instance. You should begin with the intake cycle, since intake must take place before there is anything to compress or ignite, and exhaust clears away what remains of the gasoline and air that entered the chamber in the intake cycle.

Writing process explanations is more difficult than writing descriptions or narratives. Careful ordering of steps is equally important in each, and each requires attentive selection of details. However, where unwise detail selection weakens a description or a narrative, giving too many or too few details can destroy a process explanation.

Process explanations can be made unnecessarily complicated by trying to explain a complex set of related processes as one whole when it could be explained more clearly as a set of parts. An example of this is the functioning of a central air conditioning unit. There is a natural sequence to use in explaining the operation as a whole—beginning with the fresh outside air being brought in by the outside unit and ending with the warm indoor air being returned to the outside. And if we stayed with nontechnical basics, we could easily explain the whole process by following this sequence. But it would be hopelessly confusing to explain the highly complex mechanisms involved in the compressor, condenser, control, and blower unless we described them individually. Thus, for a complete explanation of many long or complex processes, we should explain each part's function separately. Explained in stages or phases, each part's function is easily comprehensible. Then a brief overview can show how the phases fit together to form the entire process.

Two short process explanations follow. The first involves work by several people, but no equipment. The second requires people to operate the equipment, but concerns only the equipment itself.

Scheduling English Classes

Although it may seem that way to some students who cannot get the courses they want, deciding how many sections of which course to offer at what times is no hit-or-miss matter. Many hours of planning by many people are involved. A good example is the process used by the English Department.

Planning for the next semester begins in the third week of the current term. Each department receives a detailed printout on the current term's enrollment from the data processing center. Final enrollments are given for every section of every scheduled class. Classes which had to be canceled because of insufficient enrollment and those offered at the last minute because of unexpected demand are included. Similar printouts from last year are already on hand, and the data center also furnishes projected enrollment trends for the coming term.

The department chairman and secretary study the past figures and future projections to determine how many sections of each course to offer. Special attention is given to current course enrollment. For instance, six sections of technical writing were offered last term. If the average enrollment was twenty-three of a possible twenty-five students and 10 percent growth is predicted, then a seventh section can be added. However, if class size averaged only eighteen of a possible twenty-five, 10 percent predicted growth will not justify adding a section.

Once the number of sections has been determined, the remaining problem is to determine class times. Last term's figures are helpful here too. For instance, if ten o'clock classes were especially popular and eight o'clock classes were unpopular, then more ten o'clock classes and fewer eight o'clock classes may be offered. The only limitation here is classroom space. Each department can use only the number of rooms allocated to it by the academic dean. One final consideration is made for certain courses. Directors of occupational programs and athletic teams are contacted to help avoid conflicts.

When class times have been set, all that remains is to line up instructors. Each instructor gives the department chairman a proposed schedule, and the chairman tries, usually successfully, to reconcile these with the classes to be taught.

Finally, the completed schedule is sent to the data processing center and compiled with schedules from other departments, and a master schedule is printed.

The Duplicator Process

The cheapest, easiest way to make up to three hundred copies of a document is through the duplicator process. The document need only be typed or written once on a special paper containing aniline dye. This copy is generally termed the *master*. The master is attached to the revolving drum on a duplicating machine (which may be either an electric or manual model). When the "ON" lever is turned (or the hand-operated machine cranked) five things happen automatically: (1) Paper is drawn into the machine. (2) Each sheet is dampened with a liquid chemical called duplicating fluid. (3) The damp paper is pressed against the master as the drum revolves. (4) Some of the dye from the master is dissolved onto the paper, creating an exact copy. (5) The imprinted copies come out the other side and stack up in a tray.

This sketch shows how the machine operates.

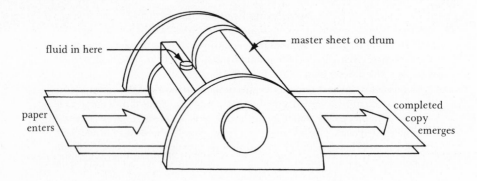

fluid in here — / master sheet on drum

paper enters / completed copy emerges

Instructions

Telling someone how to do something may seem an easy assignment, but it can be quite difficult. Just recollect the last time you tried to give someone directions on how to get somewhere; or better yet recall the last time you tried to follow someone else's directions. We assume that if we know how to do something we should be able to tell someone else how, merely by listing the steps we take. Perhaps it should be this easy, but it seldom is. In fact, nothing else that is written as often as directions is done so ineffectively.

There are several common pitfalls to avoid in giving instructions. One is writing an instruction sheet too technically, that is, writing over the reader's head. Many of the instructions for assembling children's toys would be ideal if written for engineers or experienced mechanics; but they are often unworkable for the rest of us, who merely want to put together a toy buggy or a scooter.

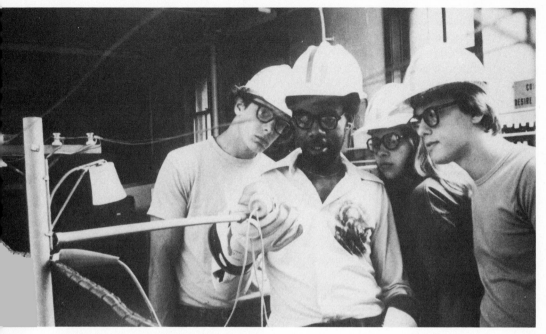

Even the best of written instructions for such complicated jobs as telephone installation frequently have to be accompanied by a personal demonstration.

Names we have never heard of are used for unfamiliar nuts, bolts, and screws. References are made to "A" or "B" on diagrams we cannot understand. Result? Either a long period of trial and error punctuated by cursing, or, if we are lucky, help from a mechanically experienced neighbor. So *rule one* is: *Define your reader and write the directions for that reader to use.* Do not just say "so simple that a child could do it;" make it that simple if you do want children to do it. If your reader is highly trained, write your directions at that level. But if your reader is someone in between—an ordinary, mechanically unskilled adult—write nearer the child's level than the expert's.

A second easily identifiable weakness of many directions is a lack of detail. Here again, failure to define the reader is at fault. Too often the writer, forgetting that the reader does not share his or her knowledge of the subject, writes what could better be termed a checklist than a workable set of instructions. He or she identifies the steps properly but only lists them instead of explaining how to do them. A common example of this error can be seen in many student-written instructions for changing a flat tire. A typical step somewhere down the line is "Next take out the jack and assemble it." This sounds good, but it would probably be useless to a reader who had never changed a tire, since assembling a jack is often one of the most difficult parts of the operation. This suggests *rule two: Always give too much rather than too little detail.* It is better to waste some of a reader's time by supplying more detail than necessary than to waste all of it by failing to give enough information for the job to be done.

Another serious pitfall in giving directions is the failure to impress upon the reader the importance of doing certain steps in a certain manner. Keep in mind this slightly changed version of what is commonly called "Murphy's law"—if something can be botched, it probably will be! In other words, the reader may make many mistakes even if you give excellent instructions. Do not give him or her an excuse to say, "I didn't know any better." If a certain time or measurement must be followed exactly, say so. For instance, many recipes include such statements as "Beat for three minutes." Often a few seconds underbeating or overbeating will make no difference; occasionally, though, beating longer than or less than the specified time will ruin the result. Warn the reader of this possibility. Another error of this type involves the apparently natural tendency of many people to tighten every nut or screw as tight as it will go. If overtightening will create problems—and it often does—warn your reader. Specify tightening the bolt exactly *x* turns, or warn of the danger of overtightening.

Although all of us could probably think of several other errors of this type, only one more need be mentioned here: failure to warn of possible physical danger. Do not rely on the reader's good common sense to warn him or her that a certain object might need to cool before being handled or that another object might be balanced precariously at a certain point. Provide a specific warning.

Rule three thus is: *Expect the worst: Warn the reader of steps that must be done especially carefully or precisely and warn of possible physical danger.*

So much for pitfalls. The *fourth* rule involves your recognizing human nature. Specifically, it is: *Never expect the reader to read the entire set of instructions before beginning step one.* Most instructions say to do just that, but numerous studies show that only one in four or five people is likely to do so even if specifically instructed to. Recognizing this fact should encourage you to build all necessary information into early steps. Any general information that the reader may need should be given before step one. If step seven is influenced by the manner in which step two is done, say so in step two rather than in step seven. Assume that the reader will plunge right in as we would probably do ourselves.

If special tools or materials are needed for the operation, tell the reader so, and list each item before step one. Nothing is more maddening than to get halfway through a job only to learn that you need additional materials or equipment and must stop to locate them. Simply prepare lists with appropriate headings, such as "materials" or "tools needed," and put them at the beginning of your explanation.

Here are a few more basic principles to guide you in your writing. *Rule five: Use the imperative, or command, point of view when telling your reader what to do.* Each actual step to be taken or move to be made should be expressed as a command: "Tighten screw *B*." "Pour in the milk." "Release the safety catch." This has nothing to do with "ordering the reader around." The imperative is simply the clearest, most concise way to give instructions. (This does not mean that every sentence must be imperative. Sentences giving explanations, warnings, or background may certainly be in some other form.)

If the operation is complicated and takes long to perform, divide the directions into several stages or phases with a few steps in each. Most operations divide logically into a certain number of phases, perhaps three or more. Look for these logical divisions and use them. Doing this has several advantages. It is easier to comprehend a set of ten steps making up a phase than a set of forty steps making up an entire operation. Also, if the reader wants to work on the operation a bit at a time or needs a few breaks, the phase divisions offer natural, easy-to-find stopping and starting points.

Use diagrams whenever you need them—but only then. Simple diagrams can often explain complex apparatus and equipment most effectively. If the apparatus is very simple and the diagram does not help the reader to do the operation, skip the diagram.

When you cannot avoid using terms unfamiliar to the reader, include a section of working definitions with only enough information to help the reader understand the instructions. Check Chapter 3 for more help with working definitions.

Three model sets of instructions follow, each written with a specific reader in mind. To use the first model, "How to Type a Spirit Master," the reader must know the fundamentals of typing. So the model uses some terminology that only someone who has had a typing course or typing experience would understand. Notice also that it merely tells the reader to clean the type, not how to clean it.

How to Type a Spirit Master

1.   Prepare your typewriter by cleaning the type thoroughly with a typecleaner or brush. Give extra attention to the broad letters such as "M" and the closed letters such as "e" and "o."

2.   Remove the tissue paper from a master set. Retain the tissue for later use. This leaves the master paper attached to the carbon sheet, with the carbon side against the master paper. When you type, the carbon surface will be facing the back side of the master paper. (The carbon sheet is usually a half-inch or so shorter than the master paper.)

3.   Insert the master paper with the attached carbon sheet into your typewriter. As the master rolls in, check to make sure that the carbon surface is facing you. (This is just the reverse of the way an ordinary carbon copy is produced.)

4.   Type the material, leaving at least a half-inch at both the top and bottom. The space at the top will be needed to insert the typed master into the duplicator. (That half-inch would not reproduce, as the carbon does not cover it.) On a manual typewriter, use sharp, even strokes, going a few words per minute slower than your normal rate. Try to type punctuation marks a little more lightly and capitals a bit more sharply.

5.   Proofread your work carefully before removing it from your typewriter. (We will explain how to make corrections later.)

6.   Remove the master paper and attached carbon from the typewriter. Replace the tissue between them to prevent smudging until you run off your copies.

The second model assumes almost nothing. If the reader can locate the circuit breaker box and tell cooled from uncooled air, he or she can follow these instructions.

Resetting Your Air Conditioner's Compressor

If you should accidentally turn your central air conditioning unit off and right back on, or if an electrical storm should cause it to kick off and on, you might find it giving only warm air. You can correct the problem quite easily by following these simple steps:

1.   Shut the unit off at the thermostat. This usually means to move the switch from "cool" to "off."

2.   Wait at least ten minutes before proceeding. This time is necessary for the unit to recycle itself. Technically, back pressure must be released. A minimum of ten minutes is needed for the recycling.

3.   Reset the circuit breaker for the air conditioning unit. Look into your electric circuit breaker box for the switch marked "air conditioner." (If the breakers are not labeled, there should be a diagram inside the door of the breaker box.) To reset the breaker, simply move the switch to "off," pause for a moment, and return it to "on."

4.   Check the outside wall of the house near your compressor for an on-off switch. Reset it as you did the circuit breaker—turn it to "off," pause for a moment, and turn it to "on."

5.   Turn the unit back on at the thermostat.

If this procedure does not result in the production of cool air, repeat the entire process, but this time wait thirty minutes for the recycling. Then if the unit still does not work properly, call a repairman.

The final model illustrates instructions designed for a very specific reader: one who has raised roses somewhere in the central or northern United States and now wants to raise them in Florida. These instructions would be of little help to a person inexperienced at growing roses, but they were not intended for one.

Planting a Rosebush in Sandy Florida Soil

1.   Prepare a hole in the usual way. Since Florida soil is predominantly sand, mix in roughly one-half peat moss to help hold in the nutrients and moisture.

2.   Purchase a Florida-grown rose with Fortuniana rootstock, preferably container grown.

3.   Remove the plant from the container, carefully checking to make sure that it is not root bound.

4.   Plant it as you would any other plant, but be certain to sprinkle a handful of any standard rose fertilizer in the top several inches of soil. After planting, make applications of fertilizer more often than you would in the North—once every four to six weeks should do.

5.   Water the rose plant daily for the first week to ten days. The sandy soil dries out very rapidly.

6.   Because rose bushes grow much larger in Florida, allow roughly twice as much distance between bushes as you would in the North.

7.   Be ready to begin your spraying or dusting program almost immediately. Insects and blackspot can appear overnight.

● EXERCISES

1. Think of a tool, small machine, or gadget that you are familiar with. Assume that you want to explain to someone how it works. Your reader is an average person, reasonably intelligent and well educated, but not familiar with the gadget. Your job is to explain how the object works; you are not writing instructions on how to operate it. Use diagrams, working definitions, or any other techniques you need.

2. Explain a process to a classmate who has no special knowledge of the process. Choose one from your work, your major field, a hobby, or use one of the suggestions blow. Remember, you just want the reader to understand the process, not how to do it.

    a. How petroleum is refined.
    b. How students are registered at your school.
    c. How votes are cast, recorded, and counted in your community.
    d. How student loans are processed.
    e. How an aquarium is set up and balanced.
    f. How satellite television works.
    g. How overdubs are done in recording.
    h. How a nautilus machine builds muscles.
    i. How hair color is changed.
    j. How a tooth is filled.

3. Select a tool or piece of equipment that you are experienced in operating. Now write a set of instructions that would enable an average person to operate the device. Assume that the reader has no special knowledge of the device. Make your instructions sufficiently foolproof and thorough that the reader can use them without reading "between the lines" or getting outside help.

4. Borrow an erector set or set of tinkertoys from a younger brother, sister, or acquaintance. Spend five minutes constructing a unique device. Now write instructions telling a classmate how to construct an identical device.

5. Write instructions telling a new student at your school how to go to the library, find a particular book, and check it out. Assume that the book is readily accessible.

6. Select a tool or piece of equipment that you are experienced in operating. Assume that your reader has used similar types or brands but not this particular one. Write instructions that he or she can use to successfully operate the tool or equipment.

5

Writing the Paper

Overview Outlines *are summaries of the organizational patterns used in developing pieces of communication.* Paragraphs, *the building blocks of longer pieces of writing, are groups of sentences expressing particular points or aspects of longer points. Paragraphs, as well as longer pieces, must be both unified and coherent.*

Once you have completed your preliminary planning, developed a thesis, and collected evidence, you have the battle more than half won. No sane person would ever claim that writing is easy, but you may be surprised at how smoothly it can go. You have four more steps to complete; and, once again, do not try to skip any. First, you will build an outline, a skeleton to flesh out with your evidence. Second, you will write a preliminary draft. Then, you will revise—and probably rewrite—the draft. Finally, you will polish it for submission. This chapter will concentrate on outlining and writing the draft. Revising and polishing will be explained in Chapter 6.

Preparing an Outline

A good outline is a structured summary of the organizational pattern you'll be using. The first step—called a *rough outline*—involves sorting and arranging your evidence into what seem to be your main points. In fact, you may have done that already while gathering your information. If not, try to sort your information into from two to five or six main topics. Any more than that would be too many for your reader to keep straight.

Assume, for instance, that you are preparing the main body of a monthly progress report on a large constructon project. Your rough outline might look like this:

Thesis: July was a very productive month, with mechanical work well ahead of schedule.
Roughed in plumbing and lavatories.
Completed masonry on west wall, fireplaces, terrace.
Rough-in electrical finished; completed east and south wings and auditorium.
Ductwork ready; vents and furnaces on site.
Walls and ceiling 50 percent completed.

Use each point in your rough outline as a major section of your detailed outline. Then fill in between with divisions and subdivisions. Consider each point in your rough outline, looking for obvious patterns. You will often be able to divide a main point into obvious divisions. For instance, you might discover that one section is a parallel chronological narrative. Your first subdivisions would name each of the parallel parts of the narrative. You could then subdivide each of these by identifying the major phases in each part. For example:

Thesis: July was a productive month, with mechanical work well ahead of schedule.
1.1 Plumbing
    1.1.1 Rough-in
    1.1.2 Finish lavatories
1.2 Masonry
    1.2.1 Outside west wall
    1.2.2 Terrace
    1.2.3 Fireplaces
1.3 Electricity
    1.3.1 Rough-in
    1.3.2 East wing
    1.3.3 South wing
    1.3.4 Auditorium
1.4 Heating and cooling
    1.4.1 Ductwork
    1.4.2 Ventilators
    1.4.3 Furnaces
1.5 Carpentry
    1.5.1 Wall partitions
    1.5.2 Ceilings

If a major section does not fit one of our basic patterns, you can subdivide it by classification or partition. A decimal outline section subdivided by classification might look like this:

5.0 Urinalysis
    5.1 Physical determinations
        5.1.1 Appearance
        5.1.2 Color
        5.1.3 Odor
        5.1.4 Volume
        5.1.5 Specific gravity
    5.2 Chemical determinations
        5.2.1 Ph
        5.2.2 Protein
        5.2.3 Ketones
        5.2.4 Bilirubin
        5.2.5 Glucose
        5.2.6 Urobilinogen
        5.2.7 Blood

Some Guidelines for Outlining

Both samples of partial outlines just shown are what are generally termed *decimal-topical outlines*. They are effective and easy to construct and to understand. The following suggestions are intended for this type of outline, but they will also help you in constructing the other types we will be discussing later.

1. Be certain that each major division is subdivided according to some pattern.

2. Make sure that you have a major division for every item you need to include.

3. Use major divisions to indicate distinct parts or sections of the document you are preparing. Do not worry if some divisions are much longer than others.

4. Do not list just one subdivision of any point. For instance, do not use 2.1 without at least a 2.2. You may use as many subdivisions as you need to, but never only one. This is not an arbitrary rule; it is a matter of simple logic. A subdivision is just that, a division or part of the whole, and a whole simply cannot be divided into one part any more than a pie can be cut into one slice. When you are tempted to use one subdivision, change the wording of the division heading. For example, if you had *1.1 Bicycles* and *1.1.1 Ten-Speed,* you could easily combine the two into *1.1 Ten-Speed Bicycles.* Using *1.1.1* Ten-Speed leads one to expect *1.1.2 Three-Speed* or some similar classification.

5. Express all subdivisions of a given point in the same grammatical form. Note the following examples:

Incorrect: 3.1 Preliminary planning
　　　　　　 3.2 The rough draft
　　　　　　 3.3 Polishing the draft

Correct: 　3.1 Planning the paper
　　　　　　 3.2 Writing the rough draft
　　　　　　 3.3 Polishing the draft

Correct: 　3.1 The preliminary plan
　　　　　　 3.2 The rough draft
　　　　　　 3.3 The polished draft

6. An outline is an aid, a crutch, not a straitjacket. Use it as a guide in putting a document together, as a framework on which to hang your material. Do not let it force you into including material you do not want to include, omitting material you do not want to omit, or placing material where you do not want it. Change your outline if you find a better organizational pattern; omit sections, combine sections, or add sections as you see fit.

Alternative Outlines

The decimal outline is fast becoming standard in most technical and professional fields; however, some fields still prefer the traditional Roman numeral outline. Complete outlines of both types are given later. As you will see from the following brief comparison, conversion from decimal to Roman numeral outline styles (or vice versa) is relatively easy:

Decimal Outline	*Roman Numeral Outline*
1.0	I
1.1	A.
1.1.1	1.
1.1.1.1	a.
2.0	II
2.1	A.
2.1.1	1.
2.1.2	2.

All of the previous examples have been of "topical outlines"; the "sentence outline" is also widely used. To construct a sentence outline, develop your patterns and rough outline, subdividing each point as usual. But instead of recording a word or phrase heading for each point and subpoint, write a good, complete sentence for each. These sentences should convey the main idea (the thesis in many cases) of each section.

Preparing a sentence outline is obviously a good deal more time consuming than developing a topic outline, but it is usually time well spent. A sentence outline not only requires you to develop organizational plans for each section of your planned document, it forces you to consider each section carefully and determine exactly what the section is supposed to communicate. A topical outline reduces time needed for revision; the sentence outline does even better. It cuts down greatly on the number of last-minute changes, additions, or deletions you will need to make.

As mentioned previously, the easy-to-construct topical-decimal outline will generally fit your needs, but you should be familiar with the Roman numeral and sentence approaches too. If an outline must be submitted as part of any complete document, you may want to check your reader's preferences. Here is a basic outline presented first in topical-decimal form, then in topical–Roman numeral form, and finally in sentence-decimal form.

Topical-Decimal Outline on the Wankel Engine

Thesis: Despite some distinct advantages over the reciprocating engine, the Wankel has never lived up to its early promise.

1.0 Early history
    1.1 Advent in 1951
    1.2 Speculation in 1960s
    1.3 Mazda's use in 1970s
2.0 Comparison with reciprocating engine
    2.1 Size
    2.2 Weight
    2.3 Number of parts
    2.4 Life expectancy
3.0 Performance
    3.1 Acceleration

  3.2 Fuel
        3.2.1 Types
        3.2.2 Amounts
4.0 Unfulfilled promise
     4.1 American automobile companies
           4.1.1 General Motors
           4.1.2 American Motors
           4.1.3 Ford and Chrysler
     4.2 Others
           4.2.1 Curtis-Wright
           4.2.2 Yammar Diesel
           4.2.3 Daimler-Benz

Topical–Roman Numeral Outline

Thesis: Despite some distinct advantages over the reciprocating engine, the Wankel has never lived up to its early promise.

 I. Early history
     A. Advent in 1951
     B. Speculation in 1960s
     C. Mazda's use in 1970's
 II. Comparison with reciprocating Engine
     A. Size
     B. Weight
     C. Number of parts
     D. Life expectancy
III. Performance
     A. Acceleration
     B. Fuel
          1. Types
          2. Amounts
IV. Unfulfilled promise
     A. American automobile companies
          1. General Motors
          2. American Motors
          3. Ford and Chrysler
     B. Curtis-Wright
          2. Yammar Diesel
          3. Daimler-Benz

Sentence-Decimal Outline

Thesis: Despite some distinct advantages over reciprocating engines, the Wankel has never lived up to its early promise.

1.0 The Wankel, or rotary, engine has existed for some thirty years, but was primarily developed in the early seventies.
     1.1 The engine was first developed in 1951 by Dr. Felix Wankel.
     1.2 During the 1960s the engine was tested and rejected by most major automobile manufacturers.

1.3 In 1972 Mazda began marketing the RX-2, the first mass-produced automobile with a rotary engine.

2.0 The rotary engine compares favorably with the piston engine on most counts.

    2.1 The Wankel is flexible enough to be made in almost any size.

    2.2 A typical Wankel weighs less than half as much as a piston engine of comparable horsepower.

    2.3 A typical Wankel has fewer than half the total parts and moving parts of a comparable piston engine.

    2.4 Depending upon the type of apex seal used, a rotary engine can last up to 500,000 miles without a major overhaul.

3.0 The Wankel consistently outperforms comparable piston engines because of its superior power-to-weight ratio, but it does not do as well on fuel economy.

    3.1 Tests show the Wankel to consistently out-accelerate piston engines of comparable horsepower.

    3.2 The rotary engine does poorly on fuel economy.

        3.2.1   It consumes more fuel.

        3.2.2   It does, however, burn a wide variety of low-priced fuels.

4.0 A few years ago many manufacturers around the world were working on rotary engines.

    4.1 American manufacturers spent the mid-seventies trying to catch up to the Japanese.

        4.1.1   GM planned to market thousands.

        4.1.2   American Motors was also going to sell GM rotaries.

        4.1.3   Ford and Chrysler had less ambitious plans.

    4.2 Other companies invested heavily in rotary development.

        4.2.1   Curtis-Wright worked for several years to meet American pollution standards.

        4.2.2   Yammar diesel worked on diesel rotaries.

        4.2.3   Daimler-Benz produced a few high-powered Wankels for use in Mercedes-Benz automobiles.

Organizing the Paper

Developing a good outline is the first step toward assuring a well-organized piece of communication. But developing effective paragraphs and making the paper unified and coherent are also crucial.

Constructing Effective Paragraphs

A paragraph is a group of sentences expressing some particular point or a particular aspect of a broader point. It is the basic organizational block of even the longest piece of writing.

In short pieces of writing, such as technical letters, each paragraph expresses a separate point and may be as short as a single sentence. But generally, paragraphs should be longer, averaging perhaps six or seven sentences. Most inexperienced technical writers mistakenly imitate the short, journalistic paragraph of the newspaper or popular magazine. Others make paragraph breaks almost randomly, sometimes running three or four distinct points into the same paragraph. Learn to develop paragraphs according to the suggestions below, and you should have no trouble with either of these problems.

The most common type of paragraph is the *expository paragraph*. Expository means explanatory—and definition, process explanation, instruction, comparison, classification, and partition are all forms of expository writing.

The basic expository paragraph typically has two levels. It begins with a general statement, usually called a topic sentence (level one), which is followed by a series of sentences furnishing specifics that clarify, exemplify, or support the topic sentence (level two). This paragraph arrangement is called *deductive* because the topic sentence comes first, followed by details. Following is a typical two-level deductive paragraph:

Topic Sentence	We were then treated to a full four-course meal, the best
Level One	Mom had ever prepared. The feast began with our favorite
	appetizer, shrimp cocktail. This was followed by a huge
	tossed salad, containing just about every vegetable growing
Level Two	in the garden. Then came the main course: roast beef,
	mashed potatoes, baby peas with onions, asparagus tips,
	and homemade rolls. As if this wasn't enough, Mom outdid
	herself by preparing each of us our favorite dessert.

Notice how the topic sentence identifies the subject—our four-course meal—and, through the key words "treated to" and "the best," comments on that meal. "Four-course" is especially important because it suggests the organization of the paragraph: four sections, one for each course. Level two consists of four parallel sentences, one for each course.

The paragraph can be converted into a three-level deductive paragraph by the addition of details about the information in the level-two sentences:

Topic Sentence	We were then treated to a full four-course meal, the best
Level Two	Mom had ever prepared. The feast began with our favorite
Level Three	appetizer, shrimp cocktail. Mom had prepared the sauce
	herself, and the shrimp were fresh from the ocean. This was
Level Two	followed by a huge tossed salad, containing just about
Level Three	every vegetable growing in the garden. The lettuce was per-
	fectly crisp and the other vegetables added just the right
Level Two	touch of flavor. Then came the main course: roast beef,
	mashed potatoes, baby peas with onions, asparagus tips,
Level Three	and homemade rolls. Every dish was done just right. The
	vegetables were fresh from the garden. The potatoes were
	swimming in melted butter. The rolls melted in your
	mouth even without butter. But the roast beef was the best.
	Done perfectly with just a touch of pink, it was the juiciest

Level Two	and tenderest I have ever tasted. As if this wasn't enough Mom outdid herself by preparing each of us our favorite
Level Three	dessert. I paid little attention to the apple pie, the Indian pudding, and the Key lime pie being eaten; I just piled more strawberries and whipped cream on my shortcake and ate away. Wow!

The entire paragraph could easily be outlined as follows:

Topic Sentence	We were then treated to a full four-course meal, the best
Level One	Mom had ever prepared.
Level Two	1.0 Appetizer
Level Three	1.1 Homemade sauce
Level Three	1.2 Fresh shrimp
etc.	2.0 Tossed salad
	2.1 Lettuce
	2.2 Other vegetables
	3.0 Main course
	3.1 Vegetables
	3.2 Potatoes and butter
	3.3 Rolls
	3.4 Roast beef
	4.0 Dessert
	4.1 Others
	4.2 My strawberry shortcake

Both two-level and three-level paragraphs are widely used. Usually your material will dictate the appropriate level. Begin with a basic two-level structure and add level-three sentences whenever needed. No particular number of level-three sentences is needed.

Checking for Unity

When we judge the organization of a piece of writing, we generally use two criteria: unity and coherence. *Unity* means oneness, relevancy to some central idea; *coherence* means fitting or sticking together. A piece of writing has unity when every section, every paragraph, every sentence relates to the same central point. Using thesis sentences and topic sentences will help ensure that you have a clear purpose in mind. Without a sharply defined purpose, you can hardly judge your composition's unity. On the other hand, if each section of your communication has a clear thesis and each paragraph has a good topic sentence, you judge unity by measuring everything against them. If something does not fit directly into the topic sentence or thesis sentence, either revise it to make it fit or remove it.

A common type of disunity in a paragraph is a sentence that fits some but not all of the elements in the topic sentence. For instance, in "Their proposal was rejected because of price and style," you are promising your reader that what follows will deal with the proposal being rejected and with price or style as determining factors. Although each of the following sentences deals with reasons why the proposal was rejected, they would violate the unity of the paragraph because they do not deal with either price or style.

1. The proposal was three weeks late arriving.
2. We were skeptical of their dependability.

If you were checking a paragraph and found such sentences, you could delete them or change your topic sentence to allow for them.

Begin checking for unity by using the thesis (central purpose) to make certain that every major section included does belong. Then do the same with each subsection and finally with each paragraph. You will sometimes have to delete material, sometimes change thesis or topic statements, and other times move items from one section or paragraph to another.

To hold these changes to a minimum, think about unity as you construct your rough draft. Develop a good, logical outline, organize each paragraph carefully, and your writing will be well unified.

Checking for Coherence

In a six-paragraph composition, unity means having all six paragraphs clearly relate to the central idea of the composition. Coherence means having Paragraph Two fit smoothly between Paragraphs One and Three and Paragraph Five fit neatly between Four and Six, and so on. Unity does not ensure coherence. In fact, many poorly constructed communications read like collections of sections all relating to a central idea but having no further relationship to one another. An incoherent piece of writing reads as though the parts were written separately then merely thrown together. It's as though you told the reader that a box of parts made up a certain machine but let him worry about how the parts fit together. Providing they are unified and otherwise well written, most incoherent compositions are readable, but they make the reader do much unnecessary work. Many readers will simply give up and quit reading. Others will read on but will be so busy trying to fit the pieces together they they will not get the full benefit of the content.

The first step toward ensuring coherence is to use a clear, logical order. No number of fancy techniques will make a set of instructions coherent if the steps are not presented in some sort of sequential order. The same is true of descriptions, of comparisons, of analyses, of any type of writing for which an effective organizational pattern is available. Even if you are merely discussing a few ideas that fit into no particular pattern, present them in some sort of logical order. You may go from most important to least important, least important to most important, most common to least common, largest to smallest, or any one of many other possible orders. Find one that makes sense; then use it.

Once you have established an effective order, you must consider transitional devices—expressions that furnish bridges or connections between the elements in an organizational pattern. Transitional devices fit the pieces together, making the communication read as a whole.

Transitional devices are needed throughout a piece of writing to tie one sentence to the next, one paragraph to the next, and one major section to the next. These devices may be a single word used as a basic part of a sentence, a single word attached to a sentence, a phrase or clause, a complete sentence, and even a complete paragraph. The most common transitional devices are discussed below.

Pronouns ● By the definition of *pronoun* you learned in grammar school—"a word that takes the place of a noun"—pronouns are obviously transitional devices. Every properly used pronoun refers to a noun used previously, so when you use a pronoun in one sentence to replace a noun used in the previous sentence, you are relating the sentences, helping to make them coherent. Since most of us use pronouns automatically, we achieve some coherence without really trying.

The catch here, of course, is that coherence is aided only by *properly used pronouns*. Improperly used pronouns ruin coherence and confuse the reader. You may want to check Chapter 16 for a refresher on pronoun usage.

You need not do anything special to use pronouns as transitional devices; merely use them properly as you would otherwise. Do not make a sentence wordy or awkward simply to work one in. Note how the following pairs of sentences are bound together by pronouns:

1. Our time cards must be turned in to our supervisor. They are then verified and sent on to the finance department.
2. Our time cards must be turned in to our supervisor. He verifies them and sends them on to the finance department.
3. Glue all connections before you quit for the day. They can set overnight and be ready for use tomorrow.
4. The patient must fast the night before the test. He or she should be given nothing at all by mouth, not even water.
5. The whole department was shocked to learn that our efforts to cut shoplifting losses had been unsuccessful. We had doubled the number of guards and had installed closed-circuit television throughout the building.

Repetition ● Using a pronoun to reflect back to a noun in a previous sentence is actually a form of repetition—the same image is expressed first in one form (as a noun) and then in another (as a pronoun). Direct repetition of the same word or expression will often give the same effect.

Although too much direct repetition sounds childish, some judiciously used repetition can be very effective in improving writing coherence. Since it is rarely used and since it is so much more obvious, repetition gives a stronger, more emphatic transition than use of pronouns. Whereas pronouns are primarily useful for transition between sentences within a paragraph, repetition of a key word or phrase is an excellent means of tying two paragraphs together. Simply use the word or phrase in the final sentence of one paragraph and repeat it in the first sentence of the next paragraph. In the examples that follow, assume that the first sentence ends one paragraph and the second starts the next:

1. Then solder the heat sinks in place.
    With the sinks firmly in place, begin the actual assembly.
2. This makes possible a total of twenty-six entirely different elevations.
    Each elevation can be used with either a three- or four-bedroom floor plan.
3. Of course, if we plant the *sinensis* hedge, we will have to remove the two specimen copper-leaf plants.
    Specimen plants are unnecessary in the back yard, as there is already an excellent small group of azaleas in one corner and a small rose garden in another.

Transitional Terms ● One very large group of words that we will call *transitional terms* is extremely helpful in providing coherence in writing. Many of these words are used automatically, just as pronouns are. But as a thoughtful writer you can use them deliberately to help bind your writing together. A partial listing of common transitional terms follows.

1. *Time words*—next, then, after, before, during, while, following, shortly, thereafter, later, on the next morning, at one o'clock, finally
2. *Place words*—over, under, above, inside, moving to the left, just beyond
3. *Contrast words*—however, but, on the other hand, to the contrary, notwithstanding, nevertheless, nonetheless
4. *Addition words*—and, furthermore, hence, moreover, likewise, similarly
5. *Cause-effect words*—therefore, so, because of this, as a result, consequently

You probably use most of these terms already. If not, familiarize yourself with each group and have a few terms of each type ready to use when needed. When using these transitional terms, it is important to remember the specific type of relationship they show. Remember too not to overuse a pet word or expression. Do not automatically use "and" when "but" would work better. "So," "as a result," and "consequently" are often preferable to "therefore." Be sure to get variety (and added interest) into your choice of transitional terms.

Besides avoiding unnecessary repetition of pet expressions, watch out for overusing the more formal expressions such as "however," "furthermore," and "moreover." Too many can make a piece of writing seem unduly formal or even pompous. Balance such words with their more common equivalents. If you must use one of these terms frequently in a particular communication, you can soften its excessive formality by sometimes putting it within a sentence rather than at the beginning. Notice how "however" can fit effectively into several places in the following sentence:

> However, we should still finish on schedule.
> We should still finish on schedule, however.
> We should, however, still finish on schedule.
> We should still, however, finish on schedule.

Notice too that when "however" is moved around within the sentence the sentence's meaning is changed slightly. When a transitional term is used within a sentence, the words immediately before and after it are given extra emphasis. So by changing the location of the term you change the emphasis of the sentence.

Transitional Sentences and Paragraphs ● Pronouns, repetition, and transitional terms can help to give coherence both within and between sentences and paragraphs. A separate transitional sentence or even a short transitional paragraph is often required to provide a connection between major sections.

Assume, for instance, that you had just said what you had to say about Lincoln Continentals and were preparing to write about Cadillacs. You would need some device to help the reader make the transition smoothly. The type of transitional aid you chose would depend upon the overall writing situation. For instance, if Continentals and Cadillacs each had a one-sentence mention in a paragraph on American-made luxury cars, almost any device—pronoun, repetition, or transitional expression—would work. If, instead, you were connecting a complete paragraph on Cadillacs with one on Continentals, a good strong repetition or transitional expression would be needed. However, if you had presented a three- or four-paragraph analysis of Continentals and were preparing a similar analysis of Cadillacs, a mere "on the other hand" or even a repetition would be too jolting. By the time your reader has read three or four paragraphs dealing specifically with one automobile, it will take more than a word or two to smoothly transfer his or her attention to another. In this case the first section of your discussion of Cadillacs could begin with a transitional sentence. You might say: "Ford has done quite well with its downsized Continentals, but General Motors has done even better with its small Cadillacs." Then you could develop your first paragraph about Cadillacs.

Even this transition might not be sufficient. For instance, say you have just completed a lengthy section on Ford automobiles and you are now going to write a similar section on General Motors automobiles. A separate paragraph relating the two could be very effective. You might use a paragraph such as the following:

> So, with its new, smaller full-sized cars hanging on to a healthy market share and its smaller luxury cars doing very well, Ford is maintaining its market share. But General Motors' new X-body is sweeping the mid-size market, its small cars are doing well, and its smaller full-size cars are still dominant. And the Cadillac is still secure in its familiar position as America's symbol of luxury. To better understand how General Motors does it, let's examine that ultimate symbol of luxury—the Cadillac.

Writing a First Draft

Almost as many writing "systems" exist as writers. The best method is the one that works best for the individual writer. Unfortunately, many writers use no method, wasting time with false starts and finally working furiously to meet a deadline. Give the method proposed here a fair try. If you develop "writer's block" try some of the alternatives suggested.

As you were answering the preliminary question "When?" you no doubt considered writing time in determining when to submit your presentation. Now go a step further and actually schedule your writing time. Make yourself sit down and write as early as possible. Allow about two hours at a sitting; you can get a good bit accomplished in that amount of time, and much more would probably be counterproductive.

To start, double-check your outline and planning worksheet. If you type, prepare to compose at the typewriter; if not, lay in a supply of paper and pencils or pens. It is always good to have your dictionary and English manual handy, but do not worry about them too much on the first draft. Your main concern is to "fill in the flesh" on your outline; you can polish the draft later.

The best place to start is usually at the beginning of the body or main section. Don't write an introduction until you have completed the material it will introduce. Try to go through your outline point-by-point. By working straight through, you will be able to keep a better perspective of the whole project than if you jumped around. Do not, however, let yourself get hung up on a section you are having trouble with; skip it and develop the next section.

Make certain that the first paragraph of each major section clearly establishes the purpose of the section with a thesis statement and that each paragraph has a clear topic sentence. Then concentrate on content. Get everything down as clearly and concisely as you can, using proper grammar, spelling, and punctuation. But do not waste time at this stage worrying about how to phrase a difficult passage, punctuate a complicated sentence, or spell an unfamiliar word. If you are in doubt about whether to include something, include it; you can always delete it later. You will find it much easier to cut down excess wordage than to build up a deficiency. Remember, you are trying to get your ideas on paper, not turn out finished copy. Once the draft is complete, put it away for a day or two.

Some other suggestions are given below:

1. Write or type in a manner that will facilitate revision. If typing, double or even triple space. If writing longhand, skip lines and leave big margins. Be neat enough that you will be able to read the draft easily.
2. If some of your information comes from outside sources, be sure to mark the text with a number coded to the source. You can add footnotes or references later.
3. If you are using illustrations, note where you will be placing them. Draw rough sketches in the text to give you better perspective.

4.   Change your outline or even your thesis if, as happens frequently, your material seems significantly different as you write it up.
5.   Use a tape recorder when you reach the "I know what I want to say, but I don't know how to say it" stage. When you play back and transcribe the tape, you will probably find that you have said it rather well. Some writers even carry cassette recorders in their automobiles to take down material as it fits together in their heads.

Persevere. Getting a draft down on paper is the hardest part, even for professional writers. Try not to procrastinate. When you get writer's block, take a break, then try again.

• CASE STUDY

Consider your present job (your most recent job if you are not now employed) or a job held by an older family member or friend. Someone in the home office has asked, through your boss, for a detailed description of your duties. You've heard rumors that your job may be upgraded, with more responsibilities and more pay. You decide to try to impress the home-office VIPs as much as possible. Develop a planning worksheet, an outline, and finally a draft of the description of your job. Feel free to invent any information you need.

• EXERCISES

1.

a.   Construct a good thesis statement for a short paper telling an acquaintance why you have selected your present major. (If you are undecided, focus on one of your possible majors.) You might write something like this: Forestry is an ideal career because it combines outdoor work with laboratory science, offers great possibilities for advancement and achievement, and provides an opportunity to help preserve our environment. Or this: Aviation technology is an appealing field of study allowing me to pursue my interests in aircraft, mechanics, and mathematics.
b.   Construct an outline in topical–Roman numeral form for a short paper (four to six paragraphs) developing your thesis.
c.   Expand the outline into sentence-decimal form.
d.   Write an introductory paragraph to set forth your thesis.
e.   Develop a paragraph for each main division (I, II, and so on) on your outline. Be sure that each has a good topic sentence.
f.   Underline the transitional devices used to connect the paragraphs.
g.   Imagine that you are writing a longer paper praising three or four major fields. Write a short transitional paragraph connecting the section you have written with a similar section (which you need not write) on another major field.
2.   Below is a scrambled outline for a research report on the aftermath of rape. Unscramble it, rearranging and rephrasing items as necessary.

The Aftermath of Rape

Thesis: An actual rape is often only the beginning, with every aspect of the victim's life affected, often for her lifetime.

    I.   Procedures immediately afterward
        A.   Police report
        B.   Examination
        C.   Don't bathe or change clothes
        D.   Types of rape
   II.   Understanding rapists' motivations
        A.   Crime of violence
        B.   The fallacy of "She was asking for it"
        C.   Usually known to victim
 III.   Effects on others
        A.   Husband
        B.   Boyfriend
        C.   Family
        D.   Friends and associates
 IV.   Stages of guilt
   V.   Steps in adjustment
        A.   Sources of help
        B.   Long-term prognosis
 VI.   Court procedures
        A.   Length of trial
        B.   Proof of crime
        C.   Tests required
            1.   Clothing
            2.   Body
        D.   Burden of proof
        E.   Defense ploys
        F.   Delay in beginning trial

3.   Locate a short article from a professional journal in your major field and prepare a topical outline of it. Make the first two sections decimal and the remainder Roman numeral. If the article is difficult to outline, be prepared to explain why.

4.   The transitional devices in the following paragraphs have been omitted. Furnish an appropriate expression for each blank.

> Good tooth brushing involves much more than merely rubbing a brush across the teeth. _____ many dentists strongly recommend using an electric toothbrush. _____ can give the proper stroking action more consistently than the human hand. _____ the gums must be brushed thoroughly. Gum disease, which almost always results from improper hygiene, causes more loss of teeth than tooth decay. _____ gum brushing will hurt at first, perhaps even bringing a little blood, but that should subside in a week or two. _____ a thorough tooth and gum brushing should last for at least five minutes. Set a timer to make sure. _____ this need only be done once a day.

5.   List all transitional devices used in the following paragraph.

> Brushing well is only the first step. Next you should floss thoroughly, getting between all teeth up to the gums. Nonwaxed, tufted floss works best. Then rinse with warm salt water, or use a water pik if you prefer. You're not done yet,

though. Chew a fluoride disclosing tablet. This will turn any remaining areas of dental plaque a deep reddish color. These areas should be brushed again. With practice you'll learn not to miss them. Remember, if you leave plaque on your teeth, decay and gum disease may occur.

6.   The following paragraphs need level-three details. Rewrite one of the paragraphs furnishing such details.

Maintaining your car's exhaust system requires caring for both the actual system and the engine. Inspect your complete exhaust system at each oil change. Keep your engine tuned, especially if you have a catalytic converter. Watch for a pungent odor or light smoke.

Proper care of oil is crucial. Change your oil at least every four thousand miles on older cars, less often on newer models. Change more often if you tow a trailer or do other heavy driving. Watch your oil light carefully.

Getting a bank loan to start a small business can be very difficult. Prepare a financial statement. Project when you will be able to pay back the loan. Be prepared to put something up as collateral.

7.   The following paragraphs lack unity. Rewrite them as well-unified paragraphs.

Paint blisters are a sign of moisture in outside walls. Water gets in back of the paint film and lifts it away from the wood siding. Severe cases may necessitate repainting. Moisture can also rot woodwork and ruin insulation. Water may also be getting through cracks or seams or around doors or windows. Blisters may also be caused by humidity vaporizing from the inside.

Loud banging in water pipes when a faucet is turned off can be very aggravating. This water hammer is a clue that additional supports are needed to prevent vibrations from opening joints and causing flooding. This flooding can be much more serious than mere noise. If your supports don't work call a plumber.

6

Revising and Polishing

Overview   *Revision should be done in several stages: content and organization must be checked, words and sentences polished, and a final copy made. A good dictionary will help you use words effectively. The most effective technical communication is written in edited English, using clear, simple, concrete words to express specific details with no deadwood.*

Completing a rough draft may be the most difficult phase of developing an important piece of communication, but revising and polishing is probably even more important. Many inexperienced writers revise and polish by recopying neatly, perhaps correcting a few minor errors. These practices are important, but much more is needed. This chapter will focus on the other qualities a polished piece of communication should possess.

A Plan for Revising

The number of revisions you make and the amount of time you spend on them will vary according to the size of the document, its importance to you, your competence, and the time available. Strangely enough, as your writing skills improve, you will probably spend more time revising. You will become more critical and will detect previously unnoticed weaknesses. And you will take the extra time to turn out work that is more than merely acceptable.

The following plan for revising and polishing important pieces of communication—give or take a step or two—should be helpful.

1.   Plan to complete your rough draft early enough to give yourself as much revision time as possible. Remember, you can spot flaws much better when you've had a day or two away from the draft.
2.   Get scissors and paste or tape to use in putting a revised draft together. You'll find many sentences, paragraphs, and short sections that merely need to be moved or rearranged. Or perhaps you'll notice an omission. A few snips and some pasting and you've saved yourself some rewriting or typing.
3.   Go through the draft, concentrating on content. Are all sections thorough enough? Is your logic sound? What about sequence? Perhaps you should move a few items. Are you sure of the facts? Mark needed changes.
4.   Read it again, this time for clarity and conciseness. Are your sentences sound? Is the vocabulary appropriate? You are bound to need a few changes here. Is the level of technicality appropriate and consistent? Point of view appropriate and consistent? Spelling? Punctuation? Mark needed changes.
5.   Put together a revised draft. Rewrite sections needing major alteration or deletion. Pencil in minor changes, and cut and paste the rest as needed.
6.   Determine final placement of all illustrations; write out references and any needed front matter, such as a title page.

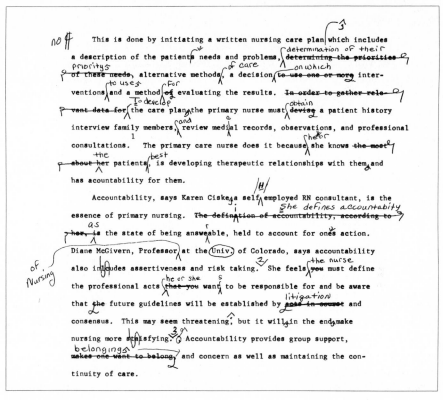

The rough draft of your communication should be examined and marked thoroughly for errors in content, organization, and grammar.

7. Give your revised draft one final check, pencilling in any changes.

8. Make your final copy, including finished illustrations. This should involve writing or typing exactly what is on your polished draft.

One particular situation should be mentioned here: some finished drafts will need to be reproduced by special means. All knowledgeable writers keep copies of every piece of communication they submit (photocopying is easier than making carbon copies). Frequently, though, important documents must be submitted in multiple copies. You will usually find this out when you are first asked to prepare the document. Ask if you are in doubt. Then determine just how the final copy is to be reproduced and what kind you will need to submit.

Choosing the Right Word

Since words are the basic medium in which we communicate, effective choice of words must rank with clear thinking among the most important components of effective communication. For that reason we will examine word usage very closely.

Using a Dictionary

In technical writing, as in general writing, the writer's best friend is the dictionary. To attempt a serious job of writing without a dictionary close by makes as little sense as trying to work trigonometry without access to trigonometric tables. Get a good desk dictionary, and do not polish a communication for submission without it. Following are some popular desk dictionaries:

The American Heritage Dictionary of the English Language. Boston: Houghton-Mifflin, 1976.

The Random House Dictionary of the English Language, College Edition. New York: Random House, 1975.

Webster's New Collegiate Dictionary. 8th ed. Springfield, Mass.: Merriam, 1975.

Webster's New World Dictionary of the American Language. Cleveland: Collins-World, 1976.

Each of these dictionaries contains information about more than 125,000 words. Any of them will meet almost all your needs. On the extremely rare occasion when you run across a word not listed, you will have to turn to an unabridged dictionary, which can be found in any library. Dictionaries less inclusive than those listed may prove inadequate.

In addition to a standard desk or collegiate dictionary, you might sometimes need access to a special dictionary of the terminology of a particular field. Such dictionaries are expensive, but they are invaluable sources of information.

Dictionaries differ considerably, both in the information they include and in their manner of presentation. Become familiar with your dictionary. Read the front matter carefully to learn how the entries are arranged. You will also find listings of the various symbols and abbreviations used. Study them too.

Below is a brief explanation of the information dictionaries generally contain, followed by a sample entry:

1.   Spelling and word division: If two or more spellings are listed, use the one given first. You will also find correct forms for plurals, possessives, comparatives, superlatives, and other inflected forms. If no form is given, assume that the word is regular—that it follows the standard form. If you have trouble finding a word, look under other ways of spelling that pronunciation, such as *ph* for *f* or *c* for *s*. The dictionary will also divide the word into syllables to show you how to divide it at the end of a line.

2.   Pronunciation: Dictionaries show the preferred pronunciation for each word. Each dictionary uses its own set of symbols (usually called *diacritical marks*). A pronunciation key is usually given at the bottom of each page to help you interpret these marks. If several pronunciations are given, the first is generally the most common, but another may be more appropriate in your geographical area.

3.   Meaning: The most important function of a dictionary is to tell us what words mean. To find the meaning you want, first determine the part

of speech. Most dictionaries give separate definitions—and sometimes separate entries—for the same word used as a noun, verb, and so on. Read the definitions given, looking for the one that best fits your context. Some dictionaries give the oldest definition first, others the most common; but the definition you need may be neither.

Remember also that dictionaries can give only a word's *denotation*—its literal, objective meaning. *Connotation*—the implied or emotive meanings conveyed by the word—must also be considered. It is discussed later in this chapter.

4. Special usage labels: Most dictionaries label words that are not standard English suitable for all types of writing and speaking. *Colloquial* (colloq.) words are more often spoken than written. *Dialect* (dial.) is more commonly used by people of a certain geographical area or social group. *Slang* words are generally considered inappropriate for formal writing. *Archaic* (arch.) and *obsolete* (obs.) words are no longer in general use. *British* words or meanings are common to England, and *foreign* words have come into English from another language. Some dictionaries even indicate a word or meaning that is specific to a particular profession. Be sure you can recognize the abbreviations for usage labels in your dictionary.

5. Synonyms: Most dictionaries list words with similar denotations and explain briefly how the words differ. Sometimes you will find a cross-reference to another entry where you can find more synonyms. Some dictionaries even include antonyms (opposites).

6. Etymology: Dictionaries also show a word's origin—its *etymology*. It is often helpful in remembering the word's meaning and in figuring out the meaning of similar words. But be careful of the etymological fallacy—the assumption that a word's original meaning has greater validity than more recent usages.

Sample Entry from *American Heritage Dictionary*

a•wait (ə-wāt′) *v.* **awaited, awaiting, awaits.** —*tr.* **1.** To wait for. **2.** To be in store for. **3.** *Obsolete.* To lie in ambush for. —*intr.* To wait. —See Synonyms at **expect.** [Middle English *awaiten,* from Old North French *awaitier,* watch for, wait on: *a-,* from Latin *ad-,* to + *waitier,* to watch, WAIT.]

Levels of Usage

American English contains many varieties. Spoken English generally differs from written English, and both vary considerably according to time and place.

These varieties have been classified in a number of ways, including such distinctions as standard versus nonstandard and formal versus informal. A reasonable beginning is to differentiate written and spoken English, which are almost never exactly the same. The sentence patterns, the words used, and the meanings attached to words differ.

Spoken English ranges from that used in casual conversation with friends or family to that used in a formal address before hundreds. Each of us has an individual *idiolect*—personal language—that differs at least a bit from those possessed by others. *Dialects*—languages spoken by groups—abound. There are regional dialects, educational-level dialects, socioeconomic-level dialects, and so on. Some dialects can be almost unintelligible to outsiders. Yet most of us can and do shift easily into a kind of bland, general dialect to communicate with any other speaker of English.

The spoken English most important on the job is *colloquial English.* It can be used in ordinary conversation, in meetings and conferences, and in any but the most formal oral presentations. Colloquial English is characterized by a relaxed adherence to the rules of grammar and syntax. It generally uses short, familiar words and occasionally includes mild slang. However, it avoids extreme forms of slang and obviously incorrect grammar and usage and can generally be labelled "correct."

Written English varies from the colloquial to the highly formal. Most lies between these extremes in what we call *edited English,* the vehicle for at least 90 percent of written business and technical communication. Edited English avoids misspellings and unconventional punctuation, mechanics, and grammar. It is conservative in usage, but not so conservative as highly formal English. Its vocabulary is quite inclusive, containing both highly technical terms and familiar, colloquial ones but avoiding long, infrequently used words and slang. It is clear, concise, and understandable to any educated person with enough technical knowledge to understand the content.

Formal English, on the other hand, adheres rigidly to traditional grammar and usage, not even permitting the use of contractions. It is also marked by a vocabulary containing many words not used in less formal English and by the absence of colloquial expressions. Sentences are generally longer and more complex.

This text is written in edited English, and the suggestions throughout are based on edited English.

Using Clear, Simple Words

One of the most common misconceptions about effective writing is that it requires long, difficult words. We mistake big words for big ideas. Instead of presenting our thoughts in the same everyday language in which they were conceived, we assume that we must somehow express them in more impressive words. Hogwash. Some few readers may be impressed by "sesquipedelian verbiage" (big words), but most would prefer to be spared the necessity of rummaging through the dictionary looking up your words and then scratching their heads trying to piece together your meaning. Some readers won't bother.

A prime contributor to the overuse of large, unfamiliar words is the misuse of a reference book that potentially is one of your best tools—the thesaurus. It is an excellent place to find exactly the right word to fit a certain context. Don't, though, use it to find more impressive synonyms for words you already know.

And don't be seduced by what scholars have termed *elegant variation*—using several elegant-sounding alternative ways of saying the same thing. Excessive repetition does sound bad, but a bit of it is preferable to using stilted words or fancy, roundabout phrasings.

One danger of trying too hard to sound impressive is the *malapropism* (often used by Norm Crosby and Archie Bunker)—the big word with the wrong meaning. Closely related to the malapropism is the big word that just doesn't fit the context. Perhaps its basic denotation fits, but it is generally used only in certain ways or in certain fields. An example is an inexperienced writer's use of *leptokurtic* to describe a group of conformists. *Leptokurtic* is a statistical term designating a group of items clustering closely around the norm; it is not generally used in other contexts.

Jargon ● One particular kind of language to avoid is *jargon.* Also called *cant* or *gobbledygook,* jargon refers to the special words and phrases used within a particular profession or field. Specific varieties of jargon have been named for their creators: legalese, bureaucratese, socioese, educationese, militarese, and the like. Some jargon is perhaps defensible as legitimate technical terminology; many technical terms have no simple English equivalents. But jargon often is marked not only by technical terminology but also by general wordiness—unnecessarily large words, too many words, very abstract words, and a roundabout, passive point of view. Such language misleads the public (although it may only intend to exclude them) by obscuring facts, inflating them, or hiding their absence. Often it merely tries to make a simple idea sound important. Consider, for instance, a phrase made popular a few years ago during the Watergate hearings, "at that point in time." How much simpler to say "then."

Below are a few examples of jargon from the field of education:

reluctant learner	lazy student
economically disadvantaged	poor
experiential approach	learning by doing
exceptional student	slow or fast learner
learning facilitator	teacher

Euphemisms ● *Euphemisms* are words or expressions that say something in a "nice way" to avoid offending someone or to make something sound better than it is. Jargon—for example, the "educationese" cited above—often employs such terms. Some euphemisms, such as those used to describe bodily functions, serve a legitimate purpose in oral communication. But even they are rarely called for in the typical technical paper. Others, such as the educationists' use of "socially maladjusted to his peer group" instead of "won't get along with his classmates," are almost indefensible when used on the general public.

Since euphemisms are almost always roundabout, as well as vague, they rob your writing of its vigor and concreteness. Don't go to the other extreme and try to offend people; just be clear and direct.

Clichés ● Another type of expression to avoid is the *cliché,* an expression that may once have had a sharp, concrete meaning but that has become stale and vague through overuse. Clichés not only fail to communicate precisely, they show a lack of concern on your part. The reader can well assume that if you used ready-made phrases instead of your own words you were equally uninspired in gathering and organizing your information. Technical writing has spawned many clichés, with the business letter perhaps the worst offender. Items are always sent "under separate cover," not "separately" or "individually." The respondent is always "in receipt of" the first letter and will respond "at a later date," "per your instructions."

The most common clichés in general writing are worn-out figures of speech such as "quick as a flash," "quiet as a mouse," or "dead as a doornail." Don't touch such expressions with a "ten-foot pole." Use your own words for your ideas.

Slang and Shoptalk ● *Slang* expressions are considered by many people to be inappropriate in serious, formal writing. Beyond that, most slang expressions quickly become clichés and may soon become dated. Those that are not yet clichéd might not be known to the reader, since slang usually begins with a small group of people and spreads slowly into general use.

Shoptalk is a special form of slang used within a given profession, company, or even office. Shoptalk often serves as a shorthand, a code, sometimes facilitating communication. A simple nickname substitutes for three or four long words in naming a machine, a tool, a process. Or a set of initials replaces the words it signifies. A voltohm milliameter becomes a VOM, or four times a day becomes QID. Shoptalk is not always bad; use it in conversation on the job, in memos, in other forms of communication with those you know well. But avoid it in communication with nontechnical people, and do not use it in formal presentations.

Neutral Words ● As mentioned earlier, words have two types of meaning. *Denotation* is the relatively objective meaning that makes a word what it is; *connotation* is the emotional or subjective meaning. If we did not understand the denotation of words, we could not use them. However, many of us use words without even thinking about their connotations.

All words—with the possible exception of *a, an, and, the*—have connotation. Some words, such as *some, any,* and *which,* have relatively little connotation, while others, such as *hippie* and *honky,* are highly charged with it. Words can have positive or negative connotation, and often the connotation is even more specific. Consider, for instance, the following group of words:

underweight	wiry
lanky	willowy
skinny	svelte
bony	trim
gaunt	slim

The denotations of the ten words vary a bit, but not much. Each could be used to describe the same person, according to how you felt toward that person or how you wanted your reader to feel. The impression conveyed could vary immensely.

None of these terms should be used in the typical job-related presentation, though. Perhaps "twenty pounds underweight" would be acceptable, but "6'1", 145 lbs." would be even better. Except in a few special situations, you should always use the most *neutral* word or expression—the one with the mildest connotation. Words with strong connotation can easily slant the meaning of a presentation, for they color the way the reader interprets your evidence. Neutral words allow the reader to draw his or her own inferences, reach his or her own conclusions, and judge yours objectively.

Another good reason to use neutral words is that they are less likely to inadvertently offend a reader. Recently, for instance, the careless use of the word *superiors* in reference to the bosses of a group of scientists and engineers brought the sharp retort "We have no superiors; we do have supervisors." Connotation, as you can see, varies from person to person. Neutral words are safer.

Building Clear, Concise Sentences

Selecting words effectively is an important step in developing a clear, concise, direct writing style. And, since words are combined into sentences as the basic building blocks of paragraphs and entire presentations, it follows that developing effective sentences is equally important. This section will help you to develop the kind of lean, meaty sentences you need. It will not help you avoid errors in grammar or sentence structure; see Chapters 15 and 16 for help there.

Words Commonly Misused

The dictionary is, of course, the best source for determining the exact spelling or meaning of a word. This list, however, contains some of the words most commonly confused or misused.

Accept–except: Accept is used as a verb meaning "to receive," while *except* is a verb meaning "to make an exception of" or "exclude." *Except* is also used as a preposition meaning "other than."

Affect–effect: As verbs, *affect* means "to change" or "influence," while *effect* means "to bring about." *Effect* is also a noun meaning "influence."

Already–all ready: Already means "by this time"; *all ready* means "completely ready."

All right: All right is often misspelled as *alright.*

A lot: A lot is often misspelled as *alot.*

Altogether–all together: Altogether is an adverb meaning "completely"; *all together* is an adjective phrase meaning "in a group."

Bad–badly: Use *bad* as an adjective (I feel bad), *badly* as an adverb (He worked badly).

Can–may: Can is used to imply ability to, *may* to imply permission to.

Complement–compliment: Complement means "to complete"; *compliment* means "to praise."

Continual–continuous: Continual implies recurring but not steady, *continuous* implies going on without interruption.

Council–counsel–consul: A *council* is a governing body; *counsel* is advice or the act of giving advice; a *consul* is a governmental official in a foreign country.

Data: Data is the plural for the rarely used singular *datum.*

Disinterested–uninterested: Disinterested means impartial; *uninterested* means not interested.

Etc.: Overused and often vague, *etc.* should be used only when the omitted items are obviously implied. It is often misspelled as *ect.*

Fewer–less: Fewer refers to number (fewer parts), *less* to amount (less fuel).

Imply–infer: Imply means "to suggest," *infer* "to conclude."

Lie–lay: Lie means "to rest" or "to recline"; *lay* means "to put or place (something) down." The past of *lie* is *lay;* the past of lay is *laid.*

Media: Media is the plural form of *medium.*

Over with–outside of: The *with* and *of* are superfluous.

Principal–principle: Principal can be an adjective (the *principal cause*) or a noun (*the principal* of a school; the *principal* plus the interest charged); *principle* is a noun meaning basic rule.

Sit–set: Set takes an object (*set* something down); *sit* does not (*sit* down).

Stationary–stationery: Stationary means fixed in position; *stationery* is writing paper.

Their–there–they're: They put on *their* uniforms. They will wait *there* for me. *They're* in a hurry.

To–too–two: To the fair; *to* work; *too* tall; do it *too; two* people.

–type: Omit *-type* in expressions like "a temperamental-*type*" person.

Whose–who's: Whose are these? *Who's* going?

–wise: Avoid jargon words such as profit*wise,* salary*wise,* or time*wise.*

Your–you're: You're going to submit *your* report.

Using Concrete Nouns and Verbs

English has often been termed a "noun and verb language," because it conveys meaning most effectively through nouns and verbs, as opposed to languages that rely more heavily on adjectives and adverbs. The most effective writers of English are therefore those who use nouns and verbs best.

The most effective nouns are the most *concrete* nouns. Concrete nouns, in this context, are nouns that zero in on the specific person, place, thing, or concept in question. *MA-2* (a respirator) is concrete; *machine* is not. *IBM Executive* (a typewriter) is concrete; *machine* is not. Nouns can be placed on what various scholars have termed a "ladder," or "staircase," of abstraction. The nearer the bottom, the more concrete the noun. Here are several examples:

less concrete	thing	food	food	automobile
more concrete	machine	vegetable	vegetable	American-made car
	electronic machine	bean	bean	Dodge
	alternator	lima	stringbean	Aspen

Notice how much more is communicated by the word at the bottom of each list than by those higher up. Concrete nouns communicate more information more concisely. Less concrete nouns communicate vague, general notions; as a result, either less gets communicated or adjectives and other modifiers are needed.

Sometimes an abstract, general noun will do. "I served food from each of the six basic food exchanges" is sufficient if the reader merely wants to know that you served a balanced meal and doesn't really need details. But *keep one point in mind:* Concrete nouns imply less concrete nouns, but the reverse is not true. "chocolate mousse" and "vanilla pudding" both imply "dessert." "Jo's Boutique" and "Saks Fifth Avenue" both indicate "clothing store." Get into the habit of using concrete nouns.

Verbs also vary in concreteness. Consider the following lists:

moving on foot	moving on foot	moving on foot
walking slowly	walking slowly	walking slowly
trudging	staggering	prancing

Like concrete nouns, concrete verbs communicate more sharply, clearly, and concisely. They reduce the need for adverbs and longer modifiers. When you read your rough draft, examine your nouns and verbs closely and be ready to substitute more concrete equivalents. Study the following pairs of sentences:

She worked on her car.	She adjusted her carburetor.
	She overhauled her engine.

He did odd jobs around the house.	He vacuumed.
	He ironed.
	He repaired a faucet.

Worked and *did* are among our weakest, most overused verbs; others are *make,* *have,* and *get.* Notice the verbs in the following sentences:

He has the job of operating a computer.
He operates a computer.

She makes use of her talents.
She uses her talents.

He gave a weak presentation of his proposal.
He presented his proposal weakly.

Giving Specific Details ● Lack of concrete language in your writing contributes to poor communication, but substituting more concrete nouns or verbs alone is insufficient. You must add more concrete *information* as well. Consider the sentences below:

The patient's wound was treated.	The nurse bandaged the patient's minor cut.
	The intern stitched the patient's eight-inch laceration.
He studied.	He wrote an English theme.
	He read his geology assignment.
	He memorized his French vocabulary.

Readers want solid, specific information. And, as pointed out earlier, even if they need only the generalities, the specific details imply the general. The specific sentences above, for instance, imply the general ones. Using specific details does not necessarily mean using a lot more words or adding extra points. In fact, straightforward details will often save words.

Using Active Voice, First-Person Constructions

Chapter 2 gave suggestions for determining when to use the active as opposed to the passive voice and when to use first-, second-, and third-person points of view. This section emphasizes that, unless content or company policy demands the passive, you should use the active voice; it also recommends using the first-person point of view whenever possible.

The passive voice almost always creates a longer sentence:

Active: Mary made up the overtime schedule.
Passive: The overtime schedule was made up by Mary.

Notice in the example that in the passive voice sentence it takes longer for us to find out who is responsible for the action that takes place. And, of course, if "by Mary" is omitted, we don't find out at all. Thus, overuse of the passive voice can—rightly or wrongly—lead to claims that the writer is being evasive. But the most common criticism of the passive voice is that it contributes to dull, slow-moving prose. No action takes place; things are merely acted upon.

Along with the passive voice, the third-person point of view is often used unnecessarily. Between the two, what could have been presented clearly and crisply becomes inflated and bland. Compare, for instance, "I tested the three samples" with "The three samples were tested by this technician." Sentences like the last one are, unfortunately, encouraged by some English teachers and some professional organizations. Why? Objectivity, they claim. Yet the reader knows who "this technician" refers to—or "this writer" or "this committee" or any other third-person substitute for "I" or "we." Any sensible reader will also recognize an objective presentation from the nature of the evidence and the way in which it is presented. When you are reporting your own efforts, use the first person unless organizational policy will not permit it.

Determining Sentence Length and Pattern

Another sometimes incorrect notion taught by many English teachers is that a good writer always strives for sentence variety, both in length and in structure. Varying sentence length is both necessary and desirable, up to a point. A steady stream of sentences all the same length certainly becomes monotonous. Further, sentences of different lengths can serve different functions. Short sentences give a brisk, emphatic effect, with long sentences slowing the reader down so that he can examine more complex, subtle matters. If you allow content to determine sentence length, you'll probably find that you have enough variety to prevent monotony, and your writing will not sound contrived.

Be particularly careful with long sentences. Don't fall into the trap of assuming that they indicate sophistication or brilliance. Extremely long sentences demand more effort from the reader and can easily become obscure. You're also more likely to make inadvertent grammatical errors in longer sentences. Twenty words or so is a good average length—some sentences much shorter, others longer.

The other commonly sought form of variety—variety in sentence type or pattern—also has some merit; but like variety in length, it can create problems. Sentence after sentence constructed in exactly the same pattern can be monotonous. Yet many professional writers rarely strive for variety. Some three-fourths of their sentences generally follow the same basic pattern—subject–verb–direct object, for example, "Samantha built the partition." (See Chapter 15 for more details on sentence patterns.) They will occasionally add an introductory clause or phrase, usually for transition. Such sentences emphasize the doer and the action. They communicate clearly and directly. Use them.

Eliminating Deadwood

Some readers are impressed by length, even demand it. They think a three-page memorandum is better than a two-pager, and so on. But most people regard excess length as a time-waster.

Cutting Down Unnecessary Clauses and Phrases

Many writers waste words and obscure meaning by using dependent clauses when a short phrase or even an adverb would suffice. Study the following examples:

She went to Cheny Tech [so that she could become an Emergency Medical Technician]. *Dependent clause*
She went to Cheney Tech to become an Emergency Medical Technician.

[While he was attending Cheney Tech,] he worked nights baking pizzas.
Dependent clause
While attending Cheney Tech, he worked nights baking pizzas.

[As is probably obvious by now,] he is highly motivated. *Dependent clause.*

Obviously, he is highly motivated.

Clauses beginning with *who, which,* or *that* can frequently be reduced to effective phrases by eliminating *who, which,* or *that* and its verb:

Problems [which were] unexpected have delayed us three weeks.
Unexpected problems have delayed us three weeks.

A candidate [who is] poorly groomed makes a bad first impression.
A poorly groomed candidate makes a bad first impression.

Enter the door [that you will see] on your left.
Enter the door on your left.

Prepositional phrases, especially those using *of* or *in,* can frequently be shortened to a single word:

Our problems are caused by the increase [in] the price [of] oil.
Our problems are caused by increased oil prices.

The starting [of] the machine is quite simple.
Starting the machine is quite simple.

Business and technical writers have made clichés of some prepositional phrases: *with (in) regard to, with (in) respect to, in (with) reference to.* Substitute *about* or *concerning:*

He wrote with regard to our Century 300 model.
He wrote about our Century 300 model.

She responded in reference to his request.
She responded concerning his request.

Determine one-word substitutes for the following clichéd phrases:

with a view to   in the event of
with hopes of   in view of
on the part of   due to the fact
            at this point in time

Sentences beginning with forms of *there is* and *it is* can almost always be cut:

There were seventeen members present.
Seventeen members were present.

It is our belief that the injuries were accidental.
We believe that the injuries were accidental.

There is only one remaining problem.
Only one problem remains.

Eliminating Throat-Clearers and Hedges

Impromptu speakers frequently use innocuous-sounding but meaningless expressions to fill space while they are deciding what to say. It sounds better than "uh, uh." Such expressions—called *throat-clearers,* for obvious reasons—have no place in writing.

Many such expressions are written, though, either through carelessness or through fruitless attempts to be diplomatic or modest. Cowardly writers often try to hedge with such expressions as *it seems to me, from my point of view, I would venture to say.* Perceptive readers can distinguish statements of fact from judgments. Repeatedly reminding them that you are stating opinion implies that you are either unsure of your opinions or afraid to commit yourself. Tell the reader you are stating opinion only when there is possible confusion between your opinion and someone else's or between fact and opinion. Here are some typical throat-clearers and hedges:

Based on my prior experience, my assessment would be. . . .
Not being totally familiar with the subject, I'd say. . . .
I'm no authority or. . . , but. . . .
Looking at it from my perspective, . . .
It is my humble judgment that. . . .

Eliminating Redundancies

Slash away redundant expressions, those that repeat themselves, say the same thing twice, as does this sentence. Here are some typical examples:
    return (revert) back
    repeat (redo, rewrite, restate) again (if you mean the second time)
    red in color
    small in size
    numerous in number

rectangular in shape
each and every
whether or not
reduce down

Other redundancies are more subtle: "We *too* decided to *also* join in." Or, "If we can avoid these *difficulties,* we shouldn't have any *trouble*." Some redundancies merely state the obvious. Consider, for instance, telling Little Bo Peep to leave alone her lost sheep that she can't tell where to find. Better yet, consider the sheep "wagging their tails *behind them*."

Closely related to redundancies are unnecessary adverbs. Some writers seem unable to use an adjective without an intensifying adverb: very best, completely satisfactory, totally empty, perfectly round, absolutely perfect. Use intensifiers only when you need to, not as a regular practice.

Making the Manuscript Sound

The typical business or technical executive is likely to be even more concerned about mechanical accuracy than the average English teacher. "Mechanical accuracy" includes spelling, punctuation, capitalization, and what most students term *grammar:* sentence structure, verb-subject agreement, pronoun reference and case, and so on. If you need help recalling—or perhaps learning—what verb form to use or where to put commas in a given sentence, check Part 3 of this text. Even if you know mechanics and grammar pretty well, familiarize yourself with the handbook section so you can refer to it as necessary.

Accurate mechanics impress the finicky reader, but they also indicate your concern about effective communication. Most readers will automatically assume that you gave the same care—or lack of it—to gathering and organizing your information as you did to checking your grammar and spelling.

More importantly, mechanical errors distract a reader and make it difficult for him or her to concentrate on what you are saying. Some errors actually cause misinterpretations. For instance, omission of the comma in the sentence "While I was taping the walls, Mort was laying carpet" does not lead to misinterpretation. But omitting the comma after the same type of clause in the sentence "While I was painting, the walls were drying" can cause confusion. Similar confusion and even serious misunderstanding can result from other mechanical problems.

Go through your revised draft carefully. Refer to a handbook for grammar, usage, and punctuation problems, and use a dictionary to check spelling. Making the final copy should be a straight typing job.

Polishing the Appearance

Some readers will not care what your document looks like. They will be satisfied if they can read it and if they like what they read. However, most readers will be strongly influenced by its appearance. Messy, sloppy-looking documents

A manual of the English language and a dictionary are invaluable resources for the writer who is polishing the revised draft.

suggest that the content is sloppy too. Neat, attractive-looking documents give the opposite impression. If you want the best possible reception for your writing, show that you care enough to make your work look good before you submit it.

Nearly everything of importance in business and industry is typed. Only informal memorandums and a very few informal reports are sometimes handwritten. If you are unsure whether to type something, type it. Typed copy looks better, is easier to read, and shows concern on your part.

Whether they are typed or handwritten, make your manuscripts neat and attractive. Margins should be straight. Keep erasures and other corrections to a minimum and make them as unobtrusive as possible. Use bond paper (onionskin is only for carbon copies). If you type, use a good, clean ribbon and keep your typewriter keys clean. Technical reports and some manuscripts are double spaced, but most documents are single spaced. Keep margins at least one inch wide.

Use good sense. Do not waste hours typing and retyping an informal memorandum to make it spotless. But a few minutes spent making your copy look attractive will pay dividends in creating a favorable impression.

A Checklist for Clarity and Conciseness

1. Have you used itemized lists? They are both clearer and more concise than paragraphs for presenting a number of similar points.

2. Have you used illustrations wisely? We've all heard that a picture is worth a thousand words, but we must also remember that one well-chosen word can be worth a thousand poorly used pictures. Chapter 7 will help you with illustrations.

3. Have you eliminated unnecessary information? Appendixes are useful places to put information that you wish to include but that isn't necessary in the document proper. See Chapter 10 for more information about appendixes.

4. Are your grammar and mechanics sound? Is the manuscript legible and neat? Errors in grammar, spelling, and punctuation are not merely signs of carelessness or lack of skill; many of them can lead to confusion or misunderstanding. Even typos can sometimes confuse or even change meaning.

5. Have you avoided going overboard in your attempt to be concise? Conciseness does not necessarily mean brevity; the shortest way may not be the clearest. Don't ruin the readability of your sentences trying to make them concise. Use conventional grammar and mechanics and good sense.

6. Have you kept your intended reader in mind? As Chapter 1 pointed out, what is clear to one reader may not be clear to another, and what is clear to the writer is not necessarily clear to the reader. When in doubt, have a friend or colleague who is similar to your intended reader check the document.

• EXERCISES

1. In the periodical section of your library, find at least one example of formal English, of edited English, and of colloquial English. Rewrite one paragraph of the colloquial English passage and one paragraph of the formal English passage in edited English.

2. Rewrite each of the following paragraphs, changing the tone as indicated:

a. Rewrite in the businesslike-but-relaxed tone:

Home computers can be of great value to the hobbyist. They can be utilized by collectors to categorize and index their collections. Other pertinent information such as transaction dates, model numbers, descriptions, and current values can be appended to such an inventory. One hobby in particular lends itself to the home computer: amateur radio operation. The hobbyist can use the computer to convert messages to Morse code, thus enabling him to send hundreds of words per minute.

b. Rewrite in the strictly business tone:

The architectural annex we propose will have enough classrooms for all our needs—no more doubling up. Also, combining McCoy Campus classes and West Campus classes will give us better coordination with the general studies programs and will be much more convenient for those of us who must commute between campuses. The proposed site is also convenient to the East Campus and to all major highways. We feel it will help to make our program.

c. Rewrite without the contentiousness:
Your promises to prospective tennants are nothing but a come-on, and you obviously had no intention of keeping them. Your maintenance crews are totally incompetent and usually foul up matters even worse, when they do bother to show up. Your pest-sprayers don't spray; your recreation directors are usually too drunk or hung over to direct; and your security guards spend too much time guarding the inside of the apartments of unattached females. But, worst of all, your little game of keep the security deposit is plain thievery.

d. Rewrite without the obsequiousness:
Please give me just a few minutes of your valuable time to answer a few questions for me. Your reputation as one of the foremost authorities in the field convinces me that you are the only one who can possibly help me in this crucial project. So, you can see that my whole project—and thus probably my job—depends upon your invaluable assistance. I know that you are very busy and have many important matters to handle, but I know you will want to help me in this time of need.

3. Rewrite the following sentences for clarity and conciseness:

a. The report which is sitting on my desk at this moment in time will be submitted by our company next week.
b. Next, lagbolts are attached to brackets A and B.
c. The report, it seems to me, is in dire need of being revised.
d. My real concern right now is whether or not the revision should be done by members of my staff or whether it might be better to send it out to someone on the outside.
e. Looking at it from my point of view, the job is too broad in scope and too time-consuming in man-hours for the chance to be taken of its not being gotten done by the prescribed deadline.
f. If we only had access to a facsimile of the report which was submitted by this department last year, the determination of what does and does not merit inclusion in this one might tend to be a bit easier.
g. The only worthwhile change he could effect that would affect the whole organization in a positive manner woulld be the tendering of his resignation.
h. He performed various operations on the machine to assure its continuing permanence.
i. He did chores around the house.
j. She drove her car up to the door and walked in.

4. Rewrite the following sentences in clear, direct English:

a. Pursuant to our previous discussion of the fourteenth, I must hereby request that your resignation be tendered to this office by the close of the day on Friday, April 12.

b.   Said resignation will be formally announced to all of those persons employed by this organization and all of those who hold subcontracts to furnish materials or expertise to said organization on or about Monday, April 15.

c.   It is hoped that you do not entertain hopes of utilizing references from this organization in seeking to procure other forms of employment elsewhere.

d.   Your job performance in performing those duties traditionally ascribed to your official job situation was exemplary; however, your attitudinal behaviors were incompatible with the desired working relationship which we continually strive to instill in those who seek to maintain membership in our leadership team.

e.   In summation, remove yourself and those belongings which belong to you immediately, as your presence in these premises is now deemed undesirable.

5.   List at least ten shoptalk terms from the same field.

6.   List at least five euphemisms and five clichés you have read or heard on television recently.

7.   Find substitutes with negative connotations for the following:

    a.   determined man
    b.   ambitious woman
    c.   unambitious woman
    d.   dependable worker
    e.   independent man

Find substitutes with positive connotations for the following:

    a.   partly cloudy
    b.   50 percent chance of rain
    c.   conniver
    d.   insubordinate employee
    e.   politician

8.   Rewrite the following sentences, using concrete verbs and nouns and furnishing specific details where possible:

    a.   He fixed the broken tool.
    b.   He had trouble with the program.
    c.   He worked on the machine.
    d.   Finally, she got it solved by using a reference book.
    e.   She had to use all her skills to get the job done.
    f.   She was well rewarded.

7

Preparing and Using Illustrations

What Illustrations Can Do

Tables

    Using Informal Tables

    Using Formal Tables

Figures

    Selecting the Right Figure

    The Bar Graph

    The Circle Graph

    The Line Graph

    The Flow Chart

    The Organizational Chart

    The Line Drawing

    Checklist for Using Figures

Exercises

Overview   Illustrations are effective in showing numerical, structural, and pictorial data; in making comparisons or contrasts; and in achieving emphasis and clarity. No great artistic skill is required to make tables, bar graphs, line graphs, circle graphs, flow charts, organizational charts, and line drawings. Knowing when and how to use these illustrations can greatly enhance your skill in technical communication.

To take advantage of illustrations in your speaking and writing, you need to know what they can do, what kinds to use in particular situations, and how to fit them into oral and written texts. If you can answer those questions, you should be able to construct illustrations to fit almost all your needs.

What Illustrations Can Do

Illustrations are excellent tools, but like any other tools, they have definite limitations. Used with care they can make a mediocre presentation more powerful; used carelessly they can ruin an otherwise effective presentation. So before you worry about how to make and use illustrations you should first get a good, clear notion of what they can and cannot do.

Add Interest ● Almost any illustration can liven up a text by offering variety, a break from page after page of type. Even the blandest table or schematic drawing provides this variety, and more elaborate illustrations, especially those in color, can provide even more. However, be careful not to use illustrations merely to add interest. Just as a funny story or other interest-getting verbal device must fit its context, so must the nonverbal device. A picture, drawing, or graph that does not fit the subject matter may simply distract readers from the rest of the text. You should particularly avoid useless illustrations in preparing highly technical material. The highly technical reader is interested in content, facts, and well-reasoned ideas. A good practice is to use illustrations to attract and maintain interest in semitechnical and especially in nontechnical material, but to make certain that each illustration is relevant and that it helps communicate in some other way too.

Give Emphasis ● Because illustrations attract attention, they automatically give emphasis to whatever they show. The reader notices them more and remembers them longer. Thus, a well-placed graph or table can give a great deal of emphasis to your key points. Make sure, though, that you put only important material in your illustrations. A beautifully constructed illustration will work against you if it misleads the reader into paying undue attention to relatively unimportant information. For this same reason, you must be careful not to use too many illustrations. The more you have, the less emphasis is given to any one.

A good illustration, used in either oral or written presentations, should attract and maintain interest as well as inform.

Illustrations can help you to emphasize certain material not just by attracting attention, but by presenting complex information in a clear, easily understood, and easily remembered form. Consider the following three presentations of the same information in a normal paragraph of type, in a table, and finally in a simple line graph. All these presentations are clear enough. You can follow the fortunes of the manufacturer in each. But the table makes the month-to-month and year-to-year fluctuations much more vivid, and the graph emphasizes them even more.

This year began with an increase as we sold 4.5 million units in January compared to 3.7 for January '79; in February we continued to improve with 4.6 compared to 3.6 for the same time in '79. March was not quite so good as we sold 4.3 against the March '79 figure of 3.9. April was about the same with 4.2 against 3.8 for April '79. May and June were slightly better with '80 figures of 4.3 and 4.4 against '79 figures of 3.7 and 3.6. We hit a slump through the next three months; in July we sold 3.9 against 3.8, in August we sold 3.8 against 3.7, and in September we sold only 3.6 against 3.8 in '79. Fortunately, sales picked up the last quarter with figures of 4.2, 4.3, and 4.6 compared with '79 sales of 3.9, 4.0, and 3.9.

Table 7.1   Comparative Sales 1979–1980

	Sales in Millions 1979	1980
Jan.	3.7	4.5
Feb.	3.6	4.6
March	3.9	4.3
April	3.8	4.2
May	3.7	4.3
June	3.6	4.4
July	3.8	3.9
Aug.	3.7	3.8
Sept.	3.8	3.6
Oct.	3.9	4.2
Nov.	4.0	4.3
Dec.	3.9	4.6

Figure 7.1 Comparative Sales for 1979–1980

Make Your Presentation More Concise ● Because your future readers will be busy people who want presentations that contain all necessary information but do not take too long to read, you must find the most concise, efficient method of presenting your material. A brief sketch of a piece of equipment, perhaps including no more than seven or eight lines and requiring less than half a page, can do the work of a full page or more of words. A good half-page table can easily contain eight to ten columns and eight to ten rows of information, perhaps as many as one hundred separate points. No matter how well you write you could not possibly say that much in the same space.

Not all illustrations are concise. A set of bar graphs or a circle graph may require more space than sentences giving the same information. Tables, drawings, and photographs, on the other hand, are almost always concise. Experiment a bit: consider conciseness, emphasis, and clarity. Sometimes you will choose the illustration, sometimes not.

Increase Clarity ● Any presentation must be clear, and illustrations often express information more clearly than words. Mathematical relationships that can be clearly visualized through graphs are almost incomprehensible to most people when presented in sentences. Drawings and photographs can show shapes and spatial relationships much more clearly than words can. Tables can present information very clearly because they line up items for comparison side-by-side, with no excessive wordage to interfere. Even simple illustrations such as bar graphs and circle graphs enhance clarity by showing things rather than merely talking about them.

Tables

Illustrations come in two general types: tables and figures. A *table* is two or more parallel columns of data. A figure thus becomes any other type of illustration. Both are widely used, and each has advantages and disadvantages. Once you decide to use an illustration you must decide whether to use a table or some sort of figure. They are frequently interchangeable; and many charts and graphs actually cannot be constructed until you have first prepared a table.

The choice between table and figure can be seen quite clearly if you look back to Table 7.1 and Figure 7.1. Both present the information clearly enough; neither is significantly more concise. The graph is a bit more emphatic and a bit more interesting, so in many cases it will be our first choice. However, the table does have some possible advantages. It can be read a bit more quickly. It is easier to construct. But most important, it is more precise. Using the numbers shown, rounded to the nearest one-hundred thousand, the graph is precise enough; but if we made our numbers more specific—for instance, 3.631 and 4.287 instead of 3.6 and 4.3—the graph would lose precision. In a table you can write down almost any number, but a graph cannot make fine distinctions. The difference between 3.7105 and 3.7085 shows up when the numbers are placed side-by-side in a table, but as contrasting heights on a graph the difference is lost. Tables also have the advantage of showing large amounts of data clearly, whereas graphs can easily become confusing if too much data is shown. This is seen clearly in Figure 7.2 and Table 7.2, in which we add data from 1976, 1977, and 1978 to the 1979 and 1980 data already shown in Figure 7.1 and Table 7.1.

Table 7.2 Comparative Sales 1976–1980					
			Sales in Millions		
	1976	1977	1978	1979	1980
Jan.	3.1	2.9	3.4	3.7	4.5
Feb.	3.0	3.0	3.5	3.6	4.6
March	3.3	3.2	3.2	3.9	4.3
April	3.6	3.4	3.3	3.8	4.2
May	3.5	3.4	3.6	3.7	4.3
June	3.5	3.4	3.7	3.6	4.4
July	3.7	3.7	3.8	3.8	3.9
Aug.	3.3	3.8	3.9	3.7	3.8
Sept.	3.2	3.7	3.6	3.6	3.6
Oct.	3.4	3.6	3.7	3.9	4.2
Nov.	3.3	3.8	3.8	4.0	4.3
Dec.	3.5	3.8	3.9	3.9	4.6

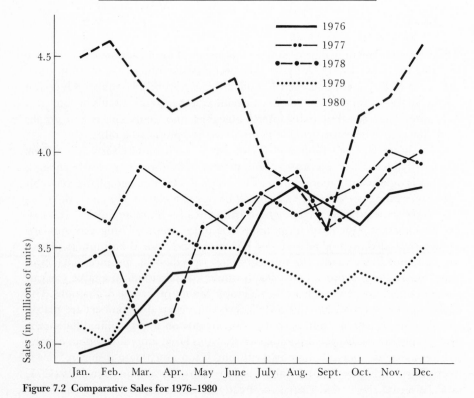

Figure 7.2 Comparative Sales for 1976–1980

In sum, use tables when you want the most concise presentation, when you want to present a large amount of information, and when precision is more important than emphasis. Another good rule of thumb is to use tables in highly technical documents and many semitechnical documents but to look for more vivid figures for less technical works.

Using Informal Tables

Tables need not be fancy. The informal table can be fitted into a text in much the same way as an ordinary paragraph. You do not even have to include it in a list of illustrations. You will sometimes have two or three short rows of information to present. Your information will be simple to read and you will not want to bother constructing a formal table. Try using an informal table like the one below:

Because of the uncertain weather picture we are going to plant fifty acres more soybeans next year that our rotational schedule calls for. We will obviously have to make some adjustments in 1982, but we cannot risk another massive corn loss for 1981.

Land	Corn	Soybeans	Other
1980 acreage	180	140	25
1981 (originally planned acreage)	140	170	35
1981 (actual acreage)	80	220	45

Notice that we are also increasing oat and clover acreage a little. We feel the changes give us a much better hedge against a third consecutive prolonged drought.

One final note of caution: use an informal table only with simple information; it should be as easy to read as any typical paragraph.

Using Formal Tables

Most tables, especially those used in reports, are formal. They are composed as separate items rather than ordinary paragraphs of text. They usually have ruled borders to set them off from the text; and if they are large enough, they may even be prepared on separate pages. Formal tables also contain descriptive titles and table numbers. They are indexed and are included in lists of illustrations. Table 7.3 shows how the information just given in the informal table would be presented in a formal table. Because the content is simple, the informal table would be more appropriate; here we merely want to see how the information would look if treated formally.

Here are some guidelines for constructing your own formal tables. You will probably find it much easier than you thought.

Table 7.3 Crop Acreages, Adcock Farms 1980–1981			
Year*	Corn	Soybeans	Other**
1980	180	140	25
1981 (original plan)	140	170	35
1981 (revised plan)	80	220	45

*All figures are in acres.   **Other includes sweet clover and oats.

1. Give every formal table a table number and title (sometimes called a *caption*). Make the title as short as possible, but give a good notion of what the table contains. Use arabic numerals (1, 2, 3) for table numbers unless you are instructed differently. You may number your tables consecutively throughout the text (Table 1, Table 2) or by chapters. The number and title are commonly centered above the table as in Table 7.3, but the style used by some organizations may call for some other placement.

2. Put a heading above each column of information to clearly indicate the contents of the column. When the items expressed are quantities, furnish the units for them in the column heading or in an explanatory note. Give all quantities in the same unit. This might require some conversion on your part, but it is better for you to do the conversion than for the reader to waste time working it out. Consider, for instance, a column giving sizes of various automobile engines. Only a very knowledgeable reader would be able to compare engine sizes given in cubic inches, cubic centimeters, and liters. A typical column heading might read "Engine Size (converted to cubic inches)."

3. Use all standard symbols and abbreviations. If your reader is highly technical you can use appropriate technical symbols and abbreviations. Just be careful not to use unfamiliar abbreviations with a nontechnical reader.

4. Use decimals to express all fractional numbers except when representing quantities normally expressed as common fractions. For instance, speak of a .625 centimeter wrench but use 7/16 inch to represent a standard American wrench.

5. If you need to present explanatory information other than that in the table proper or the title, you may use a brief legend under the title or you may use explanatory notes at the bottom of the table. For each explanatory note use either a symbol such as an asterisk (*) or a lower case letter (a) one-half space above the line referred to by the note. (See Tables 7.3 and 7.4.) Then immediately below the body of the table place the symbol or letter, again raised one-half space, and write your note. Do not confuse explanatory notes of this type with footnotes to the text. Footnotes are indicated by numbers and go at the bottom of the text.

Table 7.4 Employee Overtime for January, 1980

Employee	3	4	5	6	7	10	11	12	13	14	17	18	19	20	21	24	25	26	27	28	31	Total
Lacey	0	0	0	0	0[a]	0	0	0	0	0[a]	2	2	3	4	4	4	4	0	4	2	3	32
Applebaum	2	2	3	1	1	2	1	2	2	3	3	3	2	4	4	4	0	0	4	3	3	47
Kahn	1	1	1	2	1	2	2	1	0	0	4	6	4	4	4	4	3	4	4	0	2	50
Pyzicki	3	3	2	2	2	3	3	2	4	0	0	0	0	0	0[a]	0	0	0	0	0[a]	1	25
Chittendon	1	0	0	2	1	0	2	2	2	0	0	0	3	3	1	0	2	2	0	0	2	23
Carmody	2	2	1	0	0	2	0	0	0	2	3	3	1	0	0	3	0	0	1	1	0	21
Milke	0	1	0	0	0	1	0	2	0	0	2	0	0	1	2	0	1	0	0	0	1	12
Hunt	3	3	5	0	0	0	0	2	0	0	2	2	2	0	0	3	3	3	3	0	0	31
Anspach	0	0	0	0	0[a]	4	4	4	4	4	2	4	8[b]	0	0	8[b]	8[b]	0	0	0	0	58
Mueller	4	4	4	2	0	4	4	4	4	4	0	0	0	0	0[a]	0	0	0	0	0[a]	2	36
Totals	16	16	16	9	5	18	16	19	16	13	18	20	23	16	15	26	21	18	14	6	14	335

[a]On vacation
[b]Double Shift

6.   Construct most tables vertically so that they read like ordinary pages of print. However, you can construct them horizontally (sideways) if doing so will prevent you from running over the margins or distorting column sizes. Horizontal tables should have their titles and table numbers toward the left-hand side of the page. (See Table 7.4.)

7.   Use spacings to distinguish your rows and columns. Avoid cluttering the table with needless grid lines.

8.   After you have decided to use an illustration, have selected a formal table as the most appropriate illustration, and have determined how to construct the table, you still have one important decision: where to put it. This decision is best made in two parts. First, decide whether to put your table in the main text or to use it as an appendix. Then, if it goes into the main text, determine exactly where to put it.

If the report, or other type of document, is short and informal, your first decision is easy. Put your table in the main text and do not use appendixes. If the document is formal, base your decision on reader need. If the reader needs to see a table in order to fully comprehend the main text, put the table in the main text; you do not want the reader to have to keep referring back to an appendix. Remember, appendixes should contain supplementary material, material that may be very important but that is not needed for a good comprehension of the main text. Should you be uncertain whether to put a table in the main text or make it an appendix, include it in the main text.

When placing a formal or informal table in the text of a document, try to put it as close to the relevant textual material as you can. The ideal situation is a paragraph or two of text, the table, and more related text. If the size of the table and the nature of the text will not permit this, place the table as soon after the relevant material as you can. Never show a table before presenting the relevant text, unless both are on the same page.

You may discuss the contents of a table as much as you wish. Some reports contain one fairly short table and many pages of text discussing the implications of the table. But you need not discuss it at all. Just be sure to mention it. Even tables used as appendixes should be mentioned at an appropriate spot in the main text. You might say something like: "See Table 2.3 for complete details" or "See Appendix B."

Figures

The second category of illustration, figures, is obviously extremely varied. Anything that you can physically get into a document can be used as an illustration; and if it is not a table, it is a figure. The only limits to the variety of figures you can use are your own creativity in thinking of effective ones and your own abilities to construct various types. If you have training in art or mechanical drawing, your potential to use figures is unlimited. But even if you can barely draw a straight line, you can still construct figures to meet most of your needs. If you do have good drawing ability, this section will help you to use that ability in illustrating your written communication. If you lack drawing experience and ability, the section will show you how to construct enough basic figures to meet your needs.

Selecting the Right Figure

As we noted in the last section, figures are generally more vivid and emphatic than tables or paragraphs. They usually attract more interest than other forms of written communication, and they are much clearer than words for showing shapes, proportions, and appearances of objects or places.

The best way to get a good understanding of what simple figures can do is to consider some possible visual communication problems and see how an appropriate figure can solve each problem. Remember that no one type of figure is automatically best. Line graphs, for instance, can often be substituted for bar graphs.

1.   *Problem: How to compare amounts or sizes.* For instance, you might be looking for a powerful means of showing your financial backers how many more laboratory tests you are handling than you did last year. *Solution: A bar graph.* Bar graphs are ideal ways to compare sizes or amounts. You simply draw a series of bars, side by side, one representing each amount. The reader gets a quick, powerful impression from the differing heights of the bars. The next section will show you how to construct several types of bar graphs.

2.   *Problem: How to show graphically how some whole is divided into parts.* For instance, you might want to show your superior what portions of your work day are spent on each of your duties. *Solution: A circle graph.* Circle graphs are the standard means of showing how totals are broken down. You simply construct a circle, then slice it up as you would a pie, with the slices showing the different amounts or percentages. A later section shows a model circle graph and explains how to construct one.

3.   *Problem: How to show fluctuations during a given period of time.* For instance, you might want to show how your workload has increased during the past year. *Solution: A line graph.* Line graphs show fluctuations of any two variables, but are especially useful in showing fluctuation through time. Simply measure off time units horizontally and some other unit vertically. Plot points showing the other unit at each time period and connect the points. Increases and decreases are obvious as the line rises or falls. Helps in constructing line graphs appear later in the chapter.

4.   *Problem: How to give a clear, emphatic overview of a complicated process.* For

instance, you might want a strong, attention-getting way to summarize the basic steps in a new set of operational procedures.

Solution: A flow chart. Flow charts are often thought of as highly complex tools used by computer programmers, but they can also be simple chains of blocks used by anyone who wants to show the steps in an operation. Draw a box for each step, connect them with arrows, and write what you need to into each box. Suggestions and a model are presented in a later section.

5.   *Problem: How to show clearly a complicated organizational structure.* For instance, you might need to show new employees how your organization works, how different departments and divisions relate, and who is responsible to whom.

Solution: An organizational chart. Start at the top of the page with a small box for the head of the organization and work your way down the page with a box for every position. The reader can quickly determine who goes where and who is responsible to whom. Complete details and models are given later in the chapter.

6.   *Problem: How to show what something looks like.* For instance, you want to show a superior exactly how a proposed new workroom should be laid out. You are not an artist or a draftsman, but you do want to show what should go where.

Solution: A line drawing. Line drawings vary from simple ones done with a ruler, a compass, and a sharp pencil, to works of art done with machines costing thousands of dollars. If you can measure with a ruler and draw a straight line you can show where the various pieces of equipment should go in the proposed workroom. Suggestions are given in a later section.

The figures just mentioned will fit most communication needs of the typical technical person. Minimal equipment is needed to construct these basic illustrations. A ruler and pencil will do for most, and only a compass and protractor will be needed for the rest.

The Bar Graph

The simplest form of bar graph is a series of solid vertical bars each representing a quantity. You can construct such a graph in about ten minutes, using only paper, pencil, and ruler. Begin by selecting a scale for determining bar height. Seldom will you want to devote more than one-half page to any ordinary bar graph, so look for a scale that will make your tallest bar around four or five inches high. That should make the bars large enough so that the reader can easily recognize their differences in height. Now draw your longest bar, followed by your next longest and so on. Make each bar the same width. One-half to one inch should do nicely. If you make the bars too narrow you distort the visual impression, exaggerating the difference in bar heights; if the bars are too wide, they seem unrealistically near the same height. A good way to judge heights and widths is to use graph paper. The finished graph can then be copied or traced onto ordinary paper. Shade the bars to make them show up.

To show even more information in a bar graph, you can segment each bar. Consider, for example, a bar graph showing the number of appliances repaired by three service departments. If you could classify the types of appliances serviced into two or three categories, you could segment each bar into that number of sections. You would thus show not only comparative totals but also comparisons of each type of service operation. Construction of the segmented bar graph is essentially the same as for the ordinary bar graph. However, you now need to measure off segments as well as whole bars. You will probably also need to make the bars a bit larger so that the smaller segments will show up. To make sure that even the smallest segments do show up for easy comparison, be careful not to use too many segments. Three or sometimes four segments seem about right. Figure 7.3 shows a regular bar graph; Figure 7.4 shows a segmented one.

Both Figure 7.3 and Figure 7.4 show vertical bar graphs; however, bar graphs can also be horizontal for appropriate subjects. Vertical bars are more common because differences in height are more striking than differences in length. But if you are comparing such quantities as length or distance, horizontal bars are more natural. Construct a horizontal graph exactly as you would a vertical one, except, of course, you must measure your bars across the page. Figure 7.5 shows a typical horizontal bar graph.

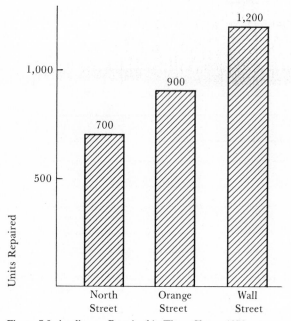

Figure 7.3 Appliances Repaired in Three Shops, 1979

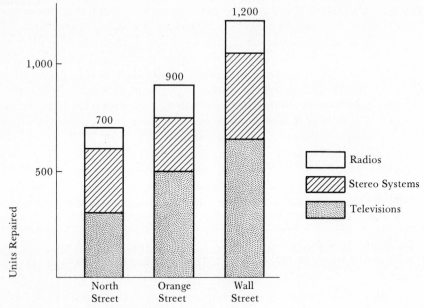

Figure 7.4 Appliances Repaired in Three Shops, 1979

Figure 7.5 Distances from Muirfield Square to Six Competitive Shopping Areas

The Circle Graph

Often called *pie charts,* circle graphs are the conventional means of showing how a total or whole is divided into parts. The pie represents the whole, and slices represent the parts. Ten or fifteen minutes should enable you to construct a good circle graph. The only tools you will need are a ruler, a pencil, a compass, and a protractor.

Begin construction by drawing a circle; one with a three- or four-inch diameter will usually do. Next, consider that the entire circle represents 100 percent. Determine the percentage represented by each part of the total. Now use your protractor to measure off 3.6° for each percent. Start at the top center and measure off your slices, beginning with the largest. Draw lines to show the slices, and the graph is complete.

To make your circle graph effective, use no more than two slices of less than 2 percent. Using a number of extremely small slices makes discrimination impossible and weakens the visual impact of the whole. If you have more than two small sections, combine them into one section called "other" and provide a list of the items included in this section, giving percentages if they will be useful. Make all lettering vertical so that the graph can be read straight on like an ordinary page of type. When the lettering needed to identify a slice will not fit properly, extend a line outside the circle and write the needed identification there. Figure 7.6 shows an effective circle graph.

The Line Graph

Bar graphs and circle graphs are all similar, but line graphs can take many different forms. They vary from simple straight lines understandable to almost anyone to complicated hyperbolas and parabolas understandable only to math-

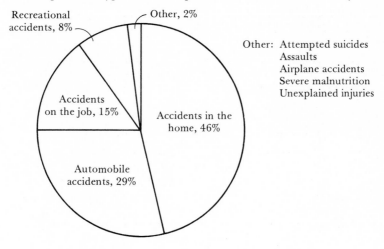

Figure 7.6 Breakdown by Types of Victims Treated at Kensington First-Aid Station

ematics majors. If you have the necessary mathematics background to construct complicated line graphs, this section will help you to use them more effectively. But our primary concern is to demonstrate how to construct ordinary, simple line graphs.

Even the simplest line graphs are based on two variables, one generally called the *controlled variable,* the other the *uncontrolled variable.* The graph shows how the uncontrolled variable fluctuates with changes in the controlled variable. For instance, you could graph the fluctuation in the boiling point of a liquid as you varied the atmospheric pressure. Pressure is the controlled variable because you decide what pressure to use. Boiling point is the uncontrolled variable because it fluctuates according to the pressure selected.

As in drawing the bar graph, you may find it easiest to construct a line graph on graph paper and then trace it onto plain paper. Begin by selecting a starting point and drawing a horizontal and a vertical line from that point. Now mark off intervals with short marks along the axes (lines) you have drawn. Put the controlled variable across the horizontal line and the uncontrolled up the vertical line. If you were trying to graph your sales figures for the past week, you would place the days of the week along the horizontal axis and a scale representing numbers of sales up the vertical axis. Now plot your values for each of the seven days and connect your points. Add a figure number and title, and you are finished. The completed graph might look like Figure 7.7.

Another excellent application of this type of line graph is to plot several lines on the same graph so that the reader may make easy comparisons. For example, we could add a line to Figure 7.7 representing the same week last year so that the reader could see if sales had improved, or we could add a line representing some other department so that the reader could see how the two departments compared. Figure 7.8 shows two lines plotted.

To make your line graphs as effective as possible, follow these guidelines:

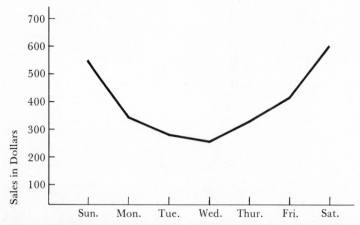

Figure 7.7 Sporting Goods Sales for September 3–9

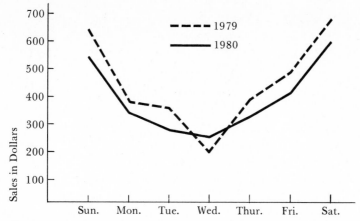

Figure 7.8 Sporting Goods Sales for First Week of September 1979, 1980

1. Pick a scale for your variables that accurately shows the relationship between the lines or points plotted. This may require some experimenting. Figure 7.9 shows how manipulating your scale can give far different visual impressions. As you can see, the variation is understated in A and overstated in C. In most cases you will want to balance the scales for your two variables as is shown in B, but use your own judgment. Select a scale that you think best shows what you want to show.

2. Remember Figure 7.2 and avoid plotting too many lines on one graph. Three is a good standard limit, but again use your own judgment. Avoid clutter and confusion.

3. Always put the controlled variable on the horizontal axis so that fluctuations appear vertically.

4. Zero is generally the basepoint where the axes meet, but it need not be. If starting at zero would bunch up the lines near the top of the graph, make some other number your base point. Try to spread the points over most of the graph whenever possible.

5. Use a key to differentiate the lines plotted. Be sure to identify the variables.

The Flow Chart

Flow charts show movement through time or space. Like line graphs, they can be so highly technical that no one without specialized training can decipher them; but fortunately, they can also be simple enough for all of us to understand. The average technical person uses them primarily to present the basic steps in complex operations. To construct a flow chart of this type, begin by constructing a series of rectangles across and down the page, one for each step in the operation. Connect the boxes with arrows indicating the proper sequence of steps. Then type or letter the names of the steps in the boxes. Figure 7.10 shows one use of a flow chart.

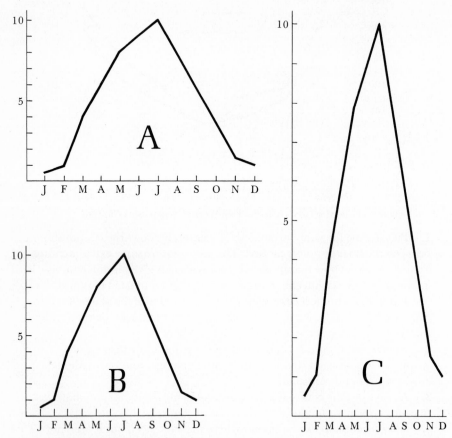

Figure 7.9 The Same Points on Three Scales

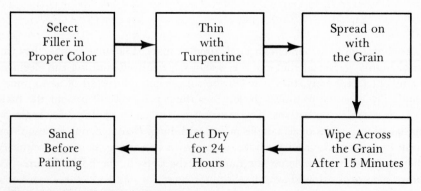

Figure 7.10 Preparing Oak for Painting

Technicians in the mechanical, electronical, and industrial fields typically use schematics and other complex drawings; those in other technical fields can generally make do with the pencil-and-ruler sketch.

The Organizational Chart

Explaining lines of authority and responsibility in large organizations is a nearly hopeless task if you rely on words alone. Fortunately, even the most complicated organizational structure can be made understandable with an organizational chart. Such a chart shows who is who and who is responsible to whom. It gives a clear, concise overview of how the organization is put together. At first glance, an organizational chart may look much like a flow chart, being composed of boxes connected by lines. However, the organizational chart is static, with lines showing relationships rather than arrows showing movement.

To construct an organizational chart, begin with a box placed in the center near the top of the page to represent the highest-ranking person in the organization. Continue by drawing solid lines down the page, with a block for every position in the chain of command. When you come to several parallel positions, use a horizontal line connected to the proper number of boxes. To show advisory or consulting relationships with no authority, use a dotted line. When you have completed your lines and boxes, place the name of the appropriate position in each box, (if you wish you can also write in the names of the individuals holding those positions). Figure 7.11 shows an effective organizational chart.

The Line Drawing

A simple ruler-and-pencil sketch of the speaker locations for a quadraphonic sound system and a detailed schematic showing the electronic structure of the entire system—both are line drawings, and both have uses as illustrations. Mechanical, electronic, and industrial technicians have to use schematics and other such complex drawings; the rest of us can generally make do with the pencil-and-ruler sketch.

121

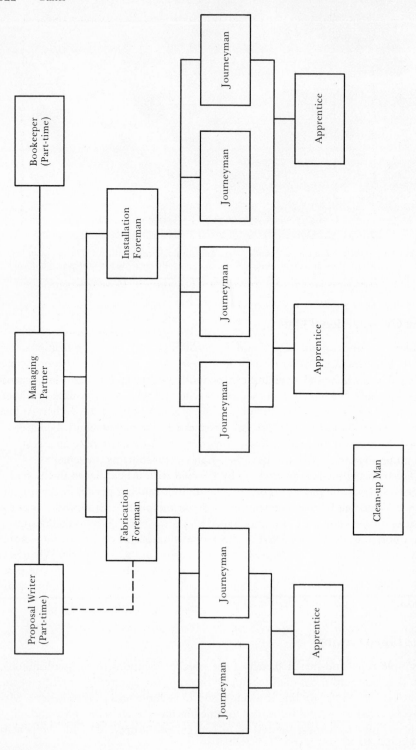

Figure 7.11 Organizational Plan of F & N Heating and Cooling

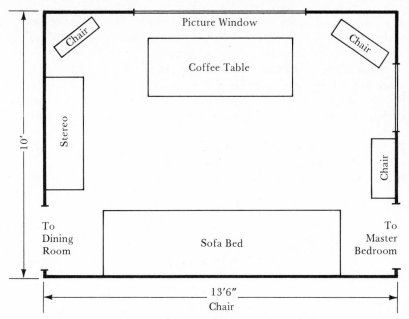

Figure 7.12 Furniture Arrangement in Living Room of Oak and Palm Models

To construct pencil-and-ruler drawings simply keep the following suggestions in mind:

1. Your work need not be beautiful, but it should be neat.
2. Everything should be drawn to some scale or at least shown in proper proportion.
3. Rectangles and squares with identifying lettering can be used to represent almost anything.
4. The fewer lines the better. Put in the drawing only the lines you want the reader to see. Do not put in extra details. Figure 7.12 shows an effective but simple line drawing.

Checklist for Using Figures

1. Put all figures in the main text unless you are certain that the reader does not need them to understand the text. Figures that are strictly supplementary can be used as appendixes.
2. Always mention a figure before it appears in the text. If you discuss the content of a figure, place your discussion as near to the figure as you can. Even if you do not discuss a figure at all, at least mention its presence.
3. Make all figures as simple and as uncluttered as you can. If you must choose between two figures, choose the simpler.

4. Do not waste space by making your figures too large. Most of them will fit nicely onto a half page or less. But make them large enough so that the differences you want to present show up.

5. Make them as neat and visually appealing as you can, particularly when you are trying to add interest or emphasis.

6. Use a figure number (Figure 1, Figure 2, or Figure 7.1, Figure 7.2) for each figure. Give each one a good descriptive title; three to eight words should do.

7. Use legends or keys whenever needed. Use a lower-case letter (a) or a symbol (*) to indicate an explanatory note.

8. Make each figure self-explanatory by using headings, legends, keys, notes, and titles. The reader should be able to understand the figure without reading the related text.

9. Use standard symbols and abbreviations but avoid those that might be too technical for your reader.

10. Make all lettering readable by a reader looking straight at the paper or turning it one-quarter turn clockwise. For example:

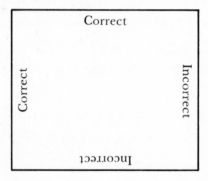

11. Do not hesitate to use figures. Look for opportunities to use them. Several of the sample reports in Part 2 of this text make effective use of illustrations.

● **EXERCISES**

1. Select a common item that you sometimes purchase, such as shampoo, sneakers, chewing gum, eight-track tapes, or cheeseburgers. Select six brands or types of the item for comparison in a formal table. Now select four qualities or characteristics to compare by. Construct a formal table showing how the six brands or types compare on the four chosen qualities.

2.

a. Determine the number of students currently enrolled in your major area at your school. Do the same for two other major areas. (You can probably exchange information with classmates here.) Construct a bar graph presenting your data.

b.   Now divide the students in each major field into first- and second-year students and third- and fourth-year students. Prepare another bar graph showing both the total number of students in each field and their breakdown into underclass or upperclass status.

c.   Determine the distance from your home to five points within or around your community. Present your data in a horizontal bar graph.

3.   Learn the total number of employees where you work (or where a parent or friend works). Divide these employees into types, using job title or type. Prepare a circle graph showing the percentage of each type.

4.   Determine the daily fluctuation over the past week of any quantity you wish. Some possibilities are high temperatures in your home town, amount of cash in your wallet, or number of clients or customers served at your place of work. Show this daily variation in a line graph. Now prepare a second graph comparing that fluctuation with the fluctuation during some other week.

5.   Select a common, simple process you use regularly, such as making a sandwich or operating your stereo system. Make sure the process involves at least four distinct steps. Construct a flow chart showing the major steps in the process.

6.   Construct a simple sketch clearly showing the furniture arrangement in one room of your home.

PART

TWO

Applications

8

Correspondence

Overview   *As you write technical letters, keep in mind the basic techniques of composition. Use the eight essential parts of a letter and other parts as needed. The all-purpose pattern—getting to the point, filling in the details, and closing smoothly—can be used in conjunction with appropriate tone and specialized techniques to meet all your technical letter-writing needs.*

The first application in Part 2 of the communication techniques discussed in Part 1 is correspondence. There are two primary reasons for beginning with this application. First, letters are required of almost everyone. Even if you do not have to write letters on the job, you will probably write some at home. Second, because most letters are shorter than other types of technical communication, they can be organized more easily, and thus they offer good practice for composing the longer, more complicated documents discussed in later chapters.

Before you begin reading about how to compose correspondence you should get several ideas clearly in mind:

1.   Plan carefully before you write even the shortest, simplest letter. If it is important enough to write, it is worth doing well.
2.   Carefully proofread, revise, and polish all letters. You are often judged as much on their form and appearance as on their content.
3.   Use the models in this chapter carefully. Each fulfills the purposes of a particular letter, but no model can fit all situations. Use the models for ideas, but do not try to twist your content to make it fit one of them.
4.   Do not hesitate to use a form slightly different from those suggested here—if you have a good reason for doing so. The forms suggested here do work; they are the most often used forms and the ones most often recommended by experts. But none are perfect, and none will work in every situation you will face. If a boss, an instructor, or your own good sense tells you to use a different form, do so.

The Letter: Form and Principles

Letters are not only the most widely written form of technical communication, they are also the most highly stylized. More books and articles have been published about the art of letter writing than about reports, procedures, instructions, and all other forms of technical writing combined. Many authorities have their own ways of writing letters, and most people who write them have their own pet authorities. Therefore, when you first begin a job that requires you to write letters, check to see if your supervisor has strong preferences about the way letters are to be written. Check also to see if there is a company style manual or guide-sheet for letter writing. If none exists, you can feel secure in writing letters exactly as this chapter recommends.

Basic Structure of the Letter

A typical well-written letter consists of the eight essential parts shown in Figure 8.1 plus certain other parts that are sometimes required and others that are optional but strongly recommended. The letter is usually written on one page, but if you have more material than you can reasonably get onto one page, you can certainly make it longer. It is almost always typed. Your instructor might accept handwritten letters in class, but if you have to send any really important letters you should try to have them typed.

The letter is set up on the page as follows:

1.   Use one-inch margins on the right and left sides. Use one and one-half inches at the bottom and one and one-half at the top if your paper does not contain a printed letterhead.
2.   Single-space the body, the inside address, and any other parts requiring more than one line.
3.   Double-space between parts such as the letterhead and dateline or salutation and body. The only exception is the spacing between the dateline and inside address. In a long letter there may be as few as four spaces here; in a short letter there may be eight or even ten.
4.   Double-space between paragraphs.

Parts of the Letter

A good technical letter must have all eight parts shown in the model and discussed below:

Letterhead ● Most organizations have letterhead stationery printed for use by their employees. If none is available and you expect to write many letters, try to get some printed. A good letterhead contains the company name, mailing address, and telephone number. It may also include a logo, or symbol of the organization. If you have to write letters without a letterhead, give only your mailing address.

Dateline ● Two spaces below the letterhead, type the dateline. You may either type it to end at the right margin or center it below the letterhead. Give the current date.

Inside Address ● Four to eight spaces below the dateline put the name and mailing address of your reader. In long letters use four spaces; in shorter letters eight or even ten. The inside address consists of three, four, or five lines, each beginning at the left margin. On the first line goes the name and, if known, the job title of the reader. On the next line goes the name of the reader's department or division, if you have this information. The third line contains the name of the organization, followed by the street address on the fourth; and the city, state, and zip code on the fifth. Double-check the inside address for accuracy. If the letter is to be typed, the typist will use this inside address on the envelope.

Letterhead {

CHURCHWELL ⚛ LABORATORIES
2918 East Graham Street   Kalamazoo, Michigan

Dateline {

July 14, 1976

(4–8 lines)

Inside Address {

Mr. George Pill
Welch Supply
3287 West 14th Street
Toledo, Ohio 42316
(2 lines)

Salutation {

Dear Mr. Pill:
(2 lines)

Body {

```
-------------------------------------------------------------
-------------------------------------------------------------
---------------------------------------

   -------------------------------------------------------
-------------------------------------------------------------
-------------------------------------------------------------
-------------------------------------------------------------
--------------------------------------------------

   -------------------------------------------------------
-------------------------------------------------------------
-------------------------------------------------
```

Complimentary Close {

(2 lines)
Yours truly,

Written Signature {

Robert O. Cox   (4–5 lines)

Signature {

Robert O. Cox
(2 lines)

Enclosure

Figure 8.1 Basic Letter Form

Salutation ● Flush to the left margin, two lines below the inside address, write the salutation. If the letter is addressed to an individual, "Dear Mr. *X*," "Dear Ms. *X*," "Dear Mrs. *X*," or "Dear Miss *X*" will do. Check past correspondence and directories to determine whether to use "Ms.," "Mrs.," or "Miss." If you know only a title such as "Chief Engineer" or "Director of Research and Development," you can use either the traditional "Dear Sir" or the now popular "Dear Sir or Madame." If you are writing to a company, a department, or some other group, you can choose between the traditional "Gentlemen" or "Gentlemen or Ladies." To avoid terms people may consider sexist, such as "Dear Sir" and "Gentlemen," without using the somewhat awkward "Dear Sir or Madame" or "Gentlemen or Ladies," try to find appropriate titles such as "Dear Director" or "Dear Associate." Avoid longer, fancier salutations, and do not use first names unless you know the reader well. Put a colon at the end of the salutation (a comma if you have used the person's first name). Some organizations now recommend omitting the salutation, but do not do so without checking.

Body ● Two lines below the salutation, begin the body. You may start each new paragraph on the left margin or after indenting five spaces. If you indent, you must also indent the first line of every other paragraph. Unless the body is extremely short (four lines or less), single-space it and double-space between paragraphs.

Complimentary Close ● Two lines below the end of the body and five spaces to the right of the center of the page, put the complimentary close. "Sincerely yours" and "Yours truly" are standard. You might occasionally use something different such as "Cordially yours," but avoid the outdated "I remain" and overly "chummy" expressions. Capitalize only the first word, and place a comma after the last word.

Signature ● Skip four spaces directly below the complimentary close and type your name. Use your business name, which may or may not include your middle name or initial. It may be followed by your title or position.

Written Signature ● Write your name longhand in the space between the complimentary close and signature. Your written signature should generally agree with the typed signature, but it does not have to. You can insert a slight personal touch by signing "Wally," or "Gloria," or "Jamie Schussel," for example.

In addition to the eight parts required in all letters, certain other parts are sometimes used:

Identification Line ● Also called the stenographic reference line, this line is typed at the left margin, two spaces below the signature. If you write your letters in longhand or type them yourself, no identification line is needed. However, if you have your letters typed for you by someone else, an identification line is required. It should consist of two sets of initials, yours in capitals followed by the typist's in lower case. It may take any of the following forms: MAV:mm, MAV/mm, MAV-mm.

Enclosure Notation ● If you enclose anything in the envelope with the letter—a check, a resume, or a sixty-page report—you must include an enclosure notation, even if you mention the enclosure in the body of the letter. When one item is enclosed, write either "Enclosure" or "Enc." When more than one item is included, use either "Encls. 2" or "Enclosures 2," noting the exact number of items. The notation is placed on the left margin two spaces below the identification line. If no identification line is used, put your enclosure notation on the left margin two spaces below the signature.

Carbon-copy Notation ● If anyone other than you and the addressee are going to receive copies of the letter, you are ethically obligated to include a carbon-copy notation. Use either the names or the titles of those receiving copies, whichever would be most meaningful to the addressee. The carbon-copy notation goes two spaces directly below the previous notation, still on the left margin. It may take any of these forms:

```
cc    Comptroller

cc:   Ms. Emily Keating

cc    1.  Mr. James Shannon
      2.  Mr. James Rahill
      3.  Ms. Norma Zielinski
```

Although not required, a third group of parts frequently used in technical letters can be quite useful in certain situations. Your careful use of them can help communicate your message more quickly and will mark you as a conscientious letter writer.

Subject Line ● A subject line does for a letter what a title does for other types of documents. It tells the reader at a glance exactly what the letter deals with. It is especially helpful in letters addressed to a company or a department because the person opening the letter can usually tell from the subject line whom to route it to. Write your subject line as you would write a title. Be brief, but give useful information. Three to eight words should do. Subject lines may be placed in several locations. The most common is two lines below the salutation and two lines above the body. If you are going to indent the paragraphs, indent the subject line accordingly. If you are not going to indent, start the subject line at the left margin. Sample subject lines follow:

Dear Ms. Iacuzzi:

    Subject: Money Count in Vault

Dear Ms. Iacuzzi:

Subject: Money Count in Vault

Attention Line ● This entry is not as popular as it once was, but you might want to use it in certain circumstances. If you think you know who should receive your letter in an organization but are not sure, use an attention line. Also use it when you prefer, but do not absolutely require, a certain person to handle your letter. After the inside address, insert the attention line two spaces below it and two spaces above the salutation, which will usually be a general one. Two sample attention lines follow:

Carbonic Industries   Carbonic Industries
180 South Court Street   180 South Court Street
Orlando, Florida 32803   Orlando, Florida 32803

Attention: Mr. Carl R. Lalumia   Attention of Mr. Carl R. Lalumia

Gentlemen: Dear Sirs:

When you use an attention line, you must also include it on the envelope. Place it on the second line of the address as shown below:

Carbonic Industries
Attention: Mr. Carl R. Lalumia
180 South Court Street
Orlando, Florida 32803

Reference Line ● A reference line tells the reader to whom a reply should be addressed. This not only saves the reader time, but it also saves you from receiving replies that must be forwarded to someone else within your organization. Type the reference line to end on the right margin. Begin it three lines below the dateline. Either of the following forms can be used:

July 27, 1981   July 27, 1981

In reply please   Reference: Ms. Jane A. Knudsen
refer to
Ms. Jane A. Knudsen

Mailing Notation ● You will often find it necessary to give special mailing instructions such as "Registered," "Certified," "Air Mail," or "Special Delivery." You might also want to give special handling instructions to the addressee. If the letter contains information that you want only the addressee to see, use "Confidential." If you wish the letter held unopened until the addressee returns to the office, you can use "Hold for Arrival." If you want the material sent to an addressee who is no longer with the organization you are writing to, use "Please Forward." Type all such notations at the left margin, midway between the dateline and the first line of the inside address. Two models are shown below:

```
REGISTERED                      HOLD FOR ARRIVAL

Mr. Robert A. Haliburton        Ms. G. R. Grazee
6142 Gateway Avenue             3114 Vista Way Drive
El Paso, Texas  71643           Honolulu, Hawaii  97871
```

Of course, for these notations to be effective, they must also be typed on the envelope. Notations to the post office are typed in capitals directly below the stamp, at least three lines above the address. Notations to the addressee are placed three lines directly under the return address. They may be typed in all capitals or may be underlined. See Figures 8.2 and 8.3.

```
Arena Trucking Co., Inc.
117 Beltline East
Topeka, Kansas  72163


HOLD FOR ARRIVAL                                         REGISTERED


               Mr. R. P. McDonald
               Acme Storage Company
               4136 East Boulware Drive
               Kansas City, MO  68714
```

Figure 8.2 Model Envelope with Mailing Notations

```
Zepher Tool and Die
818 Redwood Lane
Great Falls, Montana  89136

                                                          CERTIFIED
Confidential


               Ms. Beverly Helmer
               Doctor's Pathology Laboratory
               2214 Oswego Street
               Helena, Montana  89325
```

Figure 8.3 Model Envelope with Mailing Notations

Second-Page Heading ● If a letter requires more than one page, do not use letterhead paper for the second or for any subsequent pages. Instead, use plain paper of similar quality. Leave at least a one-inch margin at the top, then type a second-page heading; after skipping three spaces, continue the text of the letter. Either heading shown below will work.

```
Mr. Morton A. Melnick              2            February 2, 1980


will definitely create problems for us, unless we decide to contact
our members immediately.  Any telephone messages should be trans-
mitted.
```

```
Mr. Morton A. Melnick
Page 2
February 2, 1980


will definitely create problems for us, unless we decide to contact
our members immediately.  Any telephone messages should be trans-
mitted.
```

Be sure that the second page contains at least two full lines of text. If you come to the end of the first page and do not have quite enough room, either stretch the text a bit to give you the two lines of text, or shorten it so that the whole letter fits onto one page.

Envelope ● The address on an envelope must be in the right place so that the information may be read by the post office's optical character readers. Type the complete name and address single-spaced so that it is all at least one-half and no more than three inches from the bottom of the envelope. Make sure that you leave at least one inch of space from your longest line to the right-hand edge. Type the return address beginning at least one-half inch from the top and three spaces from the left edge of the envelope.

The All-Purpose Pattern

Other sections of this chapter explain several patterns for developing the bodies of specific types of letters. However, most of the letters you will write will not fit one of these patterns. More likely, your letters will require a flexible, all-purpose approach you can adopt to almost any situation. With a thorough knowledge of the specific situation, some common sense, and this basic, all-purpose pattern, you can work out a satisfactory organization for almost any letter you will have to write. This pattern is composed of three sections: (1) getting to the point; (2) filling in the details; (3) closing smoothly.

1.   Getting to the Point ● A major weakness of many letters is that their writers do not get to the point quickly enough. Explain your purpose. State your business. Seldom should you use more than one or two sentences before directly stating the basic point of the letter.

2.   Filling in the Details ● Although you can almost always introduce a letter and state its purpose in one short paragraph, you will probably need more than one paragraph to fill in the details. Your purpose will probably entail several separate points, with each requiring its own paragraph.

The manner in which you handle the details will vary as much as the types of details you have. Sometimes you will be presenting evidence. Suppose you had made a firm statement in your opening paragraph: "We can sell you three machines at $50,000 each." In filling in the details you would go on to explain that the $50,000 represented so much for labor, so much for raw material, and so forth. Almost any type of detail mentioned in Chapter 2 will be used at one time or another.

Each of the model letters following uses a slightly different type of detail, each fitting the general purpose of the letter. The paragraphs are shorter than those in most other types of writing because each makes one specific, unified, and concise point, as a good paragraph should. So do not worry about the length of your paragraphs; start a new one each time you go to a new point or a new type of detail, even if doing so makes some paragraphs only two or three sentences long.

3.   Closing Smoothly ● Your final paragraph should close the letter appropriately. Although you should usually insert some sort of goodwill note, use your common sense. If the tone of the letter has been strictly businesslike and impersonal, it would be inconsistent to say, "Let's get together for dinner some evening and discuss this matter further." However, if the letter has been friendly and warm, such a closing would be quite appropriate. Other typical goodwill endings consist of offers to be of assistance or repeated expressions of thanks.

In the model letters in this chapter, each closing is slightly different, designed to fit the tone of the first two parts of the letter. Most of them are warm and friendly, but the closings of the model claim letters on pp. 145–147 are not quite so cordial. Yet they do end the letters smoothly.

Controlling the Tone of the Letter

Chapter 2 discussed the importance of finding the appropriate tone for each piece of communication. Achieving the proper tone is crucial in writing letters. Before you begin a letter you must decide exactly what tone you want the letter to take. Should it be direct, concise, and businesslike or relaxed and "chatty"? Should it be impersonal, implying that you and the reader are communicating as representatives of two organizations? Or should it imply communication between two friends or close acquaintances? Should it be strong and assertive, or easygoing and mild? The best way to select your tone is to analyze the situation as explained in Chapters 1 and 2. Nevertheless, we can offer a few general suggestions that apply particularly to letters.

1.   Resist the temptation to be too assertive or nasty. In many letters you will be complaining to someone, telling him or her what to do, or even making demands. However, you can usually make your point clearly and emphatically without being overly aggressive. In claim letters especially resist the urge to "tell off" the company that you feel has wronged you.

2.   Do not go to the other extreme by humbling yourself or avoiding an important issue. Especially refrain from hollow flattery, which can make you sound insincere or ineffectual. (The model claim letters achieve a balanced tone—neither too aggressive nor too timid.)

3.   When dealing with the general public, try to humanize your letters. People resent being treated as numbers. Recipients of your letters should feel that they are dealing with a real person, not with an impersonal representative of a corporation or with a computer. Use goodwill endings, and in general adopt a person-to-person tone.

4.   When writing to people in important positions, be careful not to become too familiar. Keep your tone strictly businesslike. Make your goodwill ending pleasant but not overly chummy.

5.   Many regular readers of letters object to what they term the "I" point of view. This occurs when the writer constantly refers to himself or herself, especially by beginning sentences with *I*. Try to write your sentences so that *you*, rather than *I*, is stressed, thus showing the reader that you see things from his or her point of view. Employing *I* two or three times in an ordinary letter will not hurt; even starting an occasional sentence with it is all right. Just do not overdo it.

6.   Letters are one of the most frequent sources of the jargon discussed in Chapter 6. Outworn phrases such as *in reference to, pursuant to,* and *yours received as of the twenty-sixth* are used over and over. In fact, there are so many business clichés that computers have been programmed to generate entire letters from a collection of jargon phrases. As Chapter 6 makes clear, you can communicate more effectively and precisely if you avoid such phrases.

Some Common Types of Letters

If you want your letters to be as effective as possible, you should approach each one thoughtfully and methodically, doing all the preliminary planning outlined in earlier sections. This planning will lead you to establish some basic guidelines for the content and form of the letter. Then you can freely use one of the specific patterns outlined in the following pages. Remember, though, that even if one of the letter types suggested here seems to fit your circumstances exactly, you will still have to tailor the details to your situation and, especially, to your reader.

Letters of Inquiry

On occasion, you will probably need to write a letter of inquiry requesting, for example, a price list or a sales quotation or asking for some technical data from an expert in your field. If you are active in the professional world, you will sometimes need information that you can only get by written communication. How well you write the letter will often determine whether you get that information.

Letters of inquiry can be classified into two distinct types: those with reader benefit, and those without reader benefit. We will look first at the easier kind.

Letters with Reader Benefit ● If you write an equipment distributor or manufacturer asking for a catalog or price list, you will probably get a prompt answer. Distributors or manufacturers can see a possible sale as a result, so they will be willing to help you. In fact, they may do more than you ask for, sending out a salesperson to call on you. This kind of letter, because it conveys an obvious benefit to the reader, is one of the easiest types to write. But even so you must be careful about several points.

1.   If possible, begin by indicating why you are writing this particular person or organization. You can also work in a compliment. Here are two typical opening sentences:

Bill Oboe, our deputy commissioner, has been telling me for some time about your excellent. . . .

Since my arrival in San Jose, I've heard nothing but praise for Detweiler and Sons.

Be careful, though, not to overdo the flattery. You are primarily trying to get both the letter and a productive relationship off to a cordial, respectful beginning.

2.   The next sentence should lead right into the crux of the letter: the request for material or information. A sentence such as "So when we decided to enlarge our pathology lab, we naturally decided to contact you," should work nicely.

3.   Ask for what you want clearly and fully. Even if your request has obvious reader benefit, do not assume that the reader is going to send you "any information you might have about. . . ." If necessary, use several sentences to spell out exactly what you need. The only way to get precisely what you want is to ask for it. You might use simple declarative sentences in paragraph form:

I need complete data on your new word processing system. Specifications, pictures, test results, price lists, and anything else you can furnish will be appreciated. Be sure to include maintenance figures.

Or, if you want information on several specific items or points, use a numbered list of questions for clarity and convenience (see the sample letter, Figure 8.4).

4.   Be enthusiastically positive but do not make rash statements or false promises. You will probably get a quick response without having to exaggerate your needs or mislead the reader about your intentions.

5.   Close with an expression of appreciation for your reader's assistance. You can state that you look forward to receiving whatever it is you requested, but, again, do not promise anything.

A typical letter of inquiry demonstrating these general guidelines is given in Figure 8.4.

Letters without Reader Benefit. ● Much more difficult to compose than letters with obvious reader benefit are letters asking for information that the reader has no particular reason to furnish. The only obvious benefit the reader might derive is the satisfaction of doing the writer a favor. Because this frequently is an insufficient benefit, no response is made. If you need information from someone who has little reason to provide it, you will have to use all your communication skills in composing the letter of inquiry.

It has been estimated that as many as half of nonbeneficial inquiries go unanswered. Obviously, then, the following guidelines cannot guarantee you a response every time; but they can increase the odds in your favor. Study the guidelines, take a close look at the model (Figure 8.5), and be prepared to use all your persuasive skills.

1.   Since there is no self-evident benefit to the reader, try to find one. The least you can do is to show that you have a definite need for the information, that you will put it to productive use. Your reader will respond more readily if he or she knows how you are going to use the information. Telling the reader your purpose is especially important when dealing with matters of a sensitive or confidential nature.

Frequently, you can show readers that their cooperation will help you make a significant contribution to the profession. For instance, suppose you are working on a study to determine ways of keeping down fish kills in a particular lake. You write to an important ecologist asking for results of one of her recent studies. If you can show her that such information could be useful to you in examining a commonly shared problem, you are likely to get a positive response.

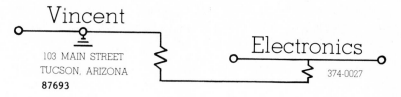

103 MAIN STREET
TUCSON, ARIZONA
87693

374-0027

December 14, 1980

Mr. Robert Frittier, Director
All Breed Dog Trainers
6418 Temple Terrace
Phoenix, Arizona   89213

Dear Mr. Frittier:

Several of my neighbors here in Monterey Industrial Park have guard dogs trained by your organization, and are most pleased with your service. Because I am planning on buying a pair of Dobermans next month, I'd like some information about your service. Could you please answer the following questions:

1. What is the charge per animal for your complete course?

2. Would I need any special facilities here for your training?

3. Would it be necessary for my assistant manager and me to participate in training in order to handle the dogs?

4. Could you double up on lessons to shorten the usual three-month training program?

My neighbors are very pleased with their dogs, and I am eager to get some of my own; so I would appreciate any information you can furnish about these questions as soon as possible.

Sincerely yours,

Gina Kaleta

Gina Kaleta
General Manager

Figure 8.4 Letter of Inquiry with Reader Benefit

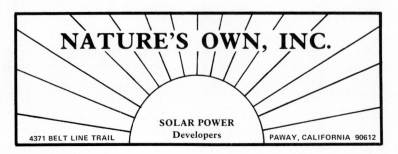

March 23, 1980

Mr. Charles Fiola, Owner
Sunny Cove Motel
6138 Atlantic Avenue
Ormond Beach, Florida 32513

Dear Mr. Fiola:

    Congratulations on your solar pool heater. Word of your inno-
vation was recently printed in the San Diego _Times_. Because we are
in the solar-heating business ourselves, my partner and I were
pleased to hear of your efforts.

    Perhaps some information exchange could help us both. We are
enclosing specifications on two pool heaters we have designed. In
return we would appreciate your answers to two short questions.

    1. What were the January and February electrical bills for
        operating the pump?

    2. What were your January and February 1978 electrical bills
        for your conventional pool heater?

    You can use the enclosed return envelope to reply, and feel
free to add any other information you wish. Even allowing for the
usual press puffery, your system seems to be the best we have seen
yet. Any suggestions you can give us will be sincerely appreciated.

                        Sincerely yours,

                        Manuel Casino

MC/sef
Enclosures 5

Figure 8.5 Letter of Inquiry without Reader Benefit

2.   Offer to share your information. If you are doing a project, study, or survey that might interest your reader, offer to send a copy of your finished product. For that matter, any meaningful proposal to return the favor will help. But do not make a vague and usually insincere-sounding statement such as, "If I can ever help you, just let me know."

3.   Formulate your questions so that they can be answered quickly and easily. Ask as few questions as you can, and make them as specific as possible. By showing your reader what you need, you may increase your chances of getting a positive response. And remember to ask for any other relevant information available.

4.   Furnish a self-addressed, stamped envelope or postcard. If possible, repeat your questions on the postcard or on a separate sheet. In short, do everything you can to make your reader's reply quick and painless.

5.   Ask the reader's permission if you plan to use the information in a published or otherwise widely circulated document.

6.   Keep your tone courteous and respectful: remember you are asking for a favor that your reader might not ordinarily have the time or inclination to grant. Do not beg or engage in false flattery. Sob stories seldom work, and insincerity almost never does.

Claim Letters

In our inefficient world, almost no one can avoid the occasional need to prepare a claim letter. An appliance that breaks down within the warranty period, a magazine subscription that expires too soon, or a mail order that does not arrive—these are some of the problems people frequently face. Sometimes the only recourse is to write a letter registering a valid claim and trying to have the trouble corrected. A claim letter then, is a formal request to get a problem resolved—to get the appliance repaired, the magazines delivered, the mail order sent.

As you know, not all claims are honored, no matter how legitimate the writers think they are. Even the best claim letter does not always achieve the desired results. In many cases, there is little reader benefit, other than building goodwill, in granting a claim. Well-written claim letters, however, do achieve results more often than poorly written ones. The guidelines below should help you write claim letters that produce positive results.

1.   Remember that your purpose is to get the claim adjusted, not to insult the person who can help you get satisfaction. Gently, but firmly, put responsibility for the problem where you think it lies. Do not make threats or wild accusations; nasty letters are self-satisfying but accomplish little else.

2.   State the problem precisely and concisely. Do not merely say "It quit working" or "It was way too big." On the other hand, do not give an overly involved analysis of the problem.

3.   Present as much evidence as you can. Refer to specific amounts, dates, and names. Enclose copies of cancelled checks, order blanks, invoices, and all other relevant documents.

4.   Suggest the way you want the adjustment made. Tell the reader, for instance, whether you wish a refund or a replacement. Make this suggestion firm but reasonable.

5.   Express confidence that the problem will be corrected. Adopt the attitude that you are pointing out a mistake the reader will be anxious to correct.

6.   If circumstances warrant it, you can point out any longstanding business relationship you may have had with the reader's organization. Doing so suggests that you might continue the relationship but is a subtle reminder that you may end it. The worst thing you can say is "You'll never get any business from me again."

7.   Use the all-purpose pattern outlined earlier. State your problem in the first one or two sentences; next, explain it more fully, provide all relevant evidence, and suggest a course of action; and finally, close by expressing confidence that the matter will be settled immediately.

8.   If you do not get satisfactory results from your first claim letter, several possible courses of action are available. You can drop the matter as a lost cause; send another letter to the same person; send a letter to the person's supervisor; or contact an attorney, a consumer-interest group, or the reader's professional organization. This decision will depend upon the amount of money involved in the claim, evidence of the claim's validity, and your willingness to pursue the matter. If you do send another letter, take a firmer stand and mention further action you might take if the claim is not resolved.

Figures 8.6 and 8.7 show initial attempts to call the reader's attention to the claim. Figure 8.8 is written to a supervisor of someone who has failed to respond to a previous letter. Note the differences in content, tone, and pattern of organization. You should freely adapt these features to the unique situations you will face.

Order Letters

Letters ordering merchandise or contracting for services are usually easy to write because your reader is obviously going to benefit from your request. Keep a few simple guidelines in mind and you can hardly go wrong.

1.   Get right to the point: State your basic order in the first sentence, without any preliminary small talk.

2.   Give clearly and fully all the information the reader will need to fill your order. Include the following:

   a.   Description or name of each article.

   b.   Number of articles of each type.

   c.   Catalog numbers (if available) for each item.

REEDY CREEK FIRE DISTRICT

LAKE BUENA VISTA, FLORIDA 32747   200-1177

June 4, 1980

Conway Enterprises
Shipping Department
13814 Lundley Street
Rockaway Center, New York   01236

Dear Sirs:

Subject: Your Order No. 63B107

On January 15 we ordered $37,021.23 worth of sprinklers and alarms from Conway Enterprises. You promised your usual one-month delivery, but the equipment still has not arrived. Please get the sprinklers to us within the next week; the alarms can wait a few extra days.

Disney Enterprises plans to open the new condominium wing in which the sprinklers will be used by May 1. Any delay in getting the sprinklers installed will cost the contractor completion time and possibly result in legal action between him and Disney.

Copies of your order form, signed and dated January 15, and your invoice dated February 7 are enclosed for your assistance in locating the cause of the delay.

In view of our eight years of excellent service from Conway Enterprises, we are confident of receiving our merchandise some-time early next week. If for some reason you cannot fulfill our request, please let us know immediately.

Sincerely yours,

Charles R. Marvin
Charles R. Marvin
Assistant Chief

CRM:lop

Enclosures 2

cc: Mr. Sean O'Hara, Chief

Figure 8.6 Model Claim Letter

Maria's Babyland
7617 Blaney Way
Salt Lake City, Utah 87807

October 24, 1981

Mr. Paul R. Brigham, General Manager
Brigham's Baby Boutique
603 Vineland Plaza
Salt Lake City, Utah 87813

Dear Mr. Brigham:

Once again you have sent us some excellent decorations for
our shop. Your merchandise is definitely the best and most
reasonably priced on the market. Unfortunately, for the third
consecutive time, you improperly filled our order.

The animal alphabet characters arrived as ordered; but the
circus-parade appliques are nowhere to be seen, and seven sets of
finger puppets arrived instead of the one set we ordered. (See
attached copies of our order and your invoice.) Please send the
circus-parade characters immediately. We are returning the six
extra sets of puppets C.O.D.

Your problem filling shipments accurately is threatening our
previously satisfactory relationship. We change our displays
regularly, so any delay in getting materials delays preparation
of the new displays and complicates our promotional plans. I
strongly urge you to make the needed changes in procedure or personnel so that we may continue to enjoy your outstanding merchandise.

We expect the circus-parade appliques within the week.

Sincerely,

Charles M. Lassiter

Charles M. Lassiter, Owner

Enclosures: 2

Figure 8.7 Model Claim Letter

III HAZELWOOD
+ GENERAL
III HOSPITAL

May 16, 1981

Ms. Winifred B. Thompson
Executive Vice-President
Buckson Industries
Lollywood Lane
Anna, Illinois 60903

It has now been more than a month since our Bucky 619-B
multivolume respirator broke down for the third time since its
installation six months ago. We have tried to contact your St.
Louis area service representative, Mr. Harold Miller, but he
has failed to respond to our request for immediate repair or
replacement of the equipment.

When the machine first broke down, seven days after we
received it, Mr. Miller repaired it within a week. The second
time, he serviced it two weeks after we contacted him. Both times
he repaired the stuck volume control and insisted that the regulator was not at fault. Now the volume control is stuck again,
and we believe the regulator is the cause. You advertise your
guarantee as the best in the field. Now is the time to prove that
claim by immediately remedying this situation.

Please see that Mr. Miller or someone else from your organization is here within ten days with a new regulator or a new
respirator. If he is not, we will be forced to purchase a similar
machine elsewhere. We are also prepared to ask the hospital legal
staff to study possible legal action if necessary and to contact
our Midwestern Association about possible censorship or blacklist-
ing of your company.

Sincerely,

Ernest T. Gary
Chief Therapist

ETG:me

1600 LINDBERGH BLVD.   HAZELWOOD, MD. 13131   (305) 831-7225

Figure 8.8 Model Claim Letter: Follow-up

    d. Price of each article.
    e. Method of payment.
    f. Total payment included or total amount to be billed.
    g. Method of shipment.
    h. Destination of shipment, if being sent to an address other than your own.
    i. Desired date of delivery.
3. Use lists rather than paragraphs if you are ordering several items.
Figure 8.9 demonstrates the correct approach.

Cover Letters

When sending out any kind of formal information, from descriptive pamphlets to lengthy governmental reports, you will need a cover letter or transmittal letter. While the terms are sometimes used interchangeably, we generally use *transmittal* to indicate letters accompanying formal documents such as major reports or proposals and *cover* to designate letters introducing other types of material. Cover and transmittal letters serve several important functions: They identify the material enclosed; they give the reason for its being sent; and they try to get the reader interested in looking at the material. Cover letters are discussed here, transmittal letters in Chapter 10.

In those cases where you are absolutely certain that your reader will read the material, you can merely describe its contents and summarize your reasons for sending it. Even when your reader has requested the information, however, you will often need to arouse his or her interest in it, a problem that requires skill in reader analysis. The writer of the model letter in Figure 8.10 provides the information requested but goes a step further by emphasizing those points in the pamphlets that he thinks the reader will be particularly interested in. This technique can be mutually beneficial; the reader gets a quick idea of the information enclosed and the writer increases the chances of selling services.

Use the following guidelines in developing your own cover letters.

1. Use a one- or two-sentence opening paragraph to identify the material enclosed and your reasons for sending it.
2. Find at least one and preferably two or three features of the enclosed material to stress in the second paragraph.
3. Mention the selected features by page number or section heading. Arouse the reader's interest enough so that he or she will actually look at the pages or sections indicated. Remember, your purpose is not to summarize the material but to get the reader to look at it.
4. In a closing sentence or two, suggest some sort of follow-up or offer to be of further assistance.

You may face a more difficult situation when you have to send copies of the same document to more than one person. In this case, you will need to analyze each of your readers and adapt the letter to their individual interests. What easily arouses your boss's curiosity, for example, may not do the same for his or her

Make your world green

1914 W. Alpine Drive   Richmond, Va. 25412
341-1172

July 17, 1981

Andover Growers' Supply
3116 Charelaton Avenue
Andover, Maryland   22134

Gentlemen or Ladies:

Please ship immediately by UPS the following supplies:

100 lbs.	Chlordane dust	$ 63.45
25 gal.	Diazinon	167.42
25 lbs.	Daconie III	118.19
10 gal.	Malathion	91.14
15 lbs.	Fore	43.17
		483.37
	4% Handling	18.93
		$502.30

I am enclosing our check for $502.30.

Sincerely yours,

Victor Benfield, Jr.

Victor Benfield, Jr.

VB:pp

Enc. Check No. 1889

Figure 8.9 Model Order Letter

ROBERT M. FRITTIER
DIRECTOR

6418 Temple Terrace
Phoenix, Ariz. 89213

December 19, 1980

Ms. Gina Kaleta
General Manager
Vincent Electronics
103 Main Street
Tucson, Arizona 89213

Dear Ms. Kaleta:

Here is the information you requested about our dog-training service.

Answers to all your specific questions are circled on the fact sheet located on page seven of the pamphlet. You might also be interested in step-by-step plans for the quick training of dog owners discussed on pages eight through twelve. Also please notice the unsolicited endorsements of our services summarized in the last section of the pamphlet.

I am convinced that our service would perfectly meet your needs, and I am anxious to discuss it with you personally. Please give me a collect call at 816-492-1673 and I'd be happy to bring a pair of Dobermans to Tucson and give you a demonstration.

Sincerely yours,

Robert M. Frittier

Robert M. Frittier
Director

RMF:dkk

Enclosure

Figure 8.10 Model Cover Letter

boss or for your organization's treasurer. What impresses one client or customer will not always impress others.

Therefore, your first task in preparing these letters is to answer your preliminary questions separately for each copy you are sending. If you are sending out too many copies for this to be practicable, try grouping your readers according to their interests, so that similar cover letters can be sent to each member of a group. Suppose, for example, that you were working in a motorcycle dealership and were planning to send out several hundred copies of a brochure on this year's new models. You certainly would not have time to compose individual cover letters. And sending copies of the same letter to all your potential customers would probably be ineffective. But you could classify the names on the mailing list into three or four groups, each likely to be interested in a particular size or type of motorcycle. Then you could compose three or four good letters. One type of letter might stress the new 250 cc. and 350 cc. Enduro models, another might mention the 1150 cc. luxury models, while a third might point out the new 150 cc. and 250 cc. trial bikes.

A Portfolio of Letters

Figures 8.11 through 8.17 present models for seven other common types of letters written by technical people. None are likely to fit your needs exactly; but if you consider them carefully along with the suggestions given earlier in this chapter, they should help you handle many letter-writing situations.

● EXERCISES

1.   Assume that you plan to attend another school upon completing your present degree program. Write a letter of inquiry to the appropriate department at a likely school. Ask at least four specific questions about matters such as programs, scholarships, transfer of credits, and admissions.

2.   Assume that you ordered a major item—one costing at least $50—several months ago. Inventing the necessary information, prepare a claim letter asking for immediate shipment.

3.   Assume that a local store has sent you a bill intended for someone else with the same last name for the third time in six months. Write a strongly worded claim letter.

4.   Write a letter ordering something that you have always wanted but could not afford.

5.   Assume that the department chairperson who responded to your letter of inquiry in Question 1 has asked for a transcript of your credits and a brief personal history. Write a cover letter to use with such material.

6.   Write a response to the claim letter in Question 4, accepting responsibility.

7.   Write a response to the claim letter in Question 4, denying responsibility.

8.   Write a letter inviting the supervisor of technical writing for some nearby large organization to come to address your class.

Buckson Industries

LOLOWOOD LANE   ANNA, ILL. 60902

Service to the Sick . . . to the Well

May 20, 1981

Mr. Ernest T. Gary, Chief Therapist
Respiratory Therapy Department
Hazelwood General Hospital
Hazelwood, Missouri 63131

Dear Mr. Gary:

Please accept my apology for your unusual difficulty with your
Bucky 619-B. Mr. Thomas Jones of my office will be in Hazelwood
next Wednesday, May 26, with a new machine.

Unfortunately, Mr. Miller, our representative in your area
left the company rather abruptly last month; and we are just now
getting his records in order. Mr. Miller's replacement, Mr. Lloyd
Fuqua, who has been in the respiratory equipment business for 18
years will be coming by with Mr. Jones to see what he can do to
restore your confidence in Buckson Industries.

I apologize for the delay and inconvenience you and your staff
have experienced and hope that our attention to this matter will
restore your confidence in our service and in our merchandise.

Sincerely yours,

Winifred B. Thompson

Winifred B. Thompson
Executive Vice-President

WBT:sjk

cc: Mr. Thomas Jones
     Mr. Lloyd Fuqua

Figure 8.11 Reply to Complaint When at Fault

Buckson Industries

LOLOWOOD LANE   ANNA, ILL. 60902

Service to the Sick . . . to the Well

May 20, 1981

Mr. Ernest T. Gary, Chief Therapist
Respiratory Therapy Department
Hazelwood General Hospital
Hazelwood, Missouri 63131

Dear Mr. Gary:

    Yes, I can certainly appreciate your dissatisfaction with a
respirator that cannot be regulated. We contacted Mr. Miller and
he agrees that your respirator has been sticking on seven.

    However, Mr. Miller is convinced that the regulator is sound
and that someone on your staff is jamming it by operating it out
of sequence. Mr. Miller and Rhoda Stamp, our electronic engineer,
will come to your department next Tuesday, May 23, to reset the
regulator. Ms. Stamp will also be available to demonstrate proper
sequencing of the controls to everyone in your department.

    As long as someone is operating the regulator improperly, a
new regulator or new respirator would not solve your problem. We
do stand behind our guarantee, and you can rest assured that we
will do everything within our power to keep your machine functioning
properly.

                        Sincerely yours,

                        Winifred B. Thompson

                        Winifred B. Thompson
                        Executive Vice-President

WBT:sjk

cc: Mr. Harold Miller
     Ms. Rhoda Stamp

Figure 8.12 Reply to Complaint When Not at Fault

Sunny Cove
Motel

6138 ATLANTIC AVE.   ORMOND BEACH, FL. 32513

- RESTAURANT
- LOUNGE
- COLOR TV
- POOL
- PLAYGROUND
- PUTTING GREEN

March 28, 1980

Mr. Manuel Casino
Nature's Own, Inc.
4371 Belt Line Trail
Paway, California 90612

Dear Mr. Casino:

Thank you for the very complimentary letter about my pool-heating
project. I am very excited about my system and about the future
of solar heating. It is gratifying to know that organizations such
as Nature's Own are engaging in the same sort of work.

Unfortunately, my attorney, Carla R. LaCorte, has temporarily for-
bidden me to give out any information about my system. She expects
to have patent work completed by the first of May, after which I
will not only answer your questions but will furnish complete
information about our system.

I will be contacting you again as soon as I am able. Meanwhile,
good luck in your own efforts, and thanks for your material.

Sincerely,

Charles Fiola

Charles Fiola

CF:kjk

cc: Ms. Carla R. LaCorte

Figure 8.13 Negative Reply to Inquiry

Sunny Cove Motel

6138 ATLANTIC AVE.   ORMOND BEACH, FL. 32513

- RESTAURANT
- LOUNGE
- COLOR TV
- POOL
- PLAYGROUND
- PUTTING GREEN

March 28, 1980

Mr. Manuel Casino
Nature's Own, Inc.
4371 Belt Line Trail
Paway, California 90612

Dear Mr. Casino:

Our January electric bill went down from $271.13 to $118.67 and our February bill from $381.71 to $219.69, a bit below my projected savings but not bad considering the unusually rainy weather we've been experiencing.

Enclosed are complete specifications on my system and copies of some research reports I've put together in developing and refining the system.

Congratulations on your work. If enough of us keep working and sharing our knowledge, we can beat this energy problem yet.

Cordially,

Charles Fiola

Charles Fiola

CF: ljf

Enclosures: 4

Figure 8.14 Positive Reply to Inquiry

May 23, 1980

Dr. Roseanne Dana
Department of Nutrition
Shands Teaching Hospital
Gainesville, Florida 31907

Dear Dr. Dana:

   Congratulations on your recent series in <u>Today's Health</u>, and
thank you for your common-sense analysis of the many myths about
so-called miracle foods and vitamins. It is heartwarming to see
someone with your credentials saying the same things we've been
trying to tell the public for several years.

   Because of this and your many other accomplishments in public
nutrition, we members of the Duvall County Nutritional Association
would like you to be guest of honor at our annual banquet to be
held at seven-thirty at the Airport Inn on Thursday, June 21. We
would like you to speak to use for twenty to thirty minutes, giving
any pointers you can on how to convince the public of the need for
balanced nutrition.

   As a self-supporting group of forty-five working nutritionists,
we cannot afford to offer you an honorarium; but we will, of course,
cover your expenses for the drive and give you a well-balanced,
delicious meal.

   Please drop us a brief letter of acceptance by June 7 so that
we can complete preparations for our best banquet ever. Our entire
membership is looking forward to thanking you in person for your
outstanding service to our cause.

                         Sincerely,

                         Martin Heddison

                         Martin Heddison

Figure 8.15 Letter Extending an Invitation

Shands Teaching Hospital

May 27, 1980

Mr. Martin Heddison
Highpoint Nutritional Services
367 Highpoint Plaza
Jacksonville, Florida   31716

Dear Mr. Heddison:

Thank you for your gracious invitation and kinds words of praise. Unfortunately, I must decline the offer to speak to your group, since I will be in New York that weekend attending a seminar. Please do invite me again, however; I am anxious to meet the members of your organization and to do all I can to help you in your very important work. Perhaps we could get together some evening during July or August.

Meanwhile, let me recommend Dr. Linda Jo Christiansen, my associate here at Shands. Dr. Christiansen has recently completed some significant metabolic studies, and you could hear her rather startling findings before they hit the journals. Contact Dr. Christiansen here, and I'm certain that she will be delighted to cooperate.

Let me once again express my regrets about the schedule conflict. Please invite me again soon.

Cordially,

Dr. Roseanne Dana

Dr. Roseanne Dana

RD/pgc

Figure 8.16 Letter Rejecting an Invitation

June 22, 1980

Dr. Linda Jo Christiansen
Department of Nutrition
Shands Teaching Hospital
Gainesville, Florida 31907

Dear Dr. Christiansen:

Thanks! You did more last night to boost the morale of the
Duvall County Nutritional Association and to give us direction in
our work than we can ever thank you for. I came to work this
morning feeling good about myself and my profession and with re-
newed confidence that the battle against charlatans and misguided
saviors of our nation's health can be won. Your reception last
night should certainly show that my colleagues agree.

Thank you again, and congratulations on your successful
research. The people of North Florida are lucky to have you serving
them.

Respectfully,

Martin Heddison

Martin Heddison

Figure 8.17 Thank-You Letter

9

Memorandums

Overview   *The memorandum is the basic medium of communication within organizations. It can be a brief note or a short, informal report. Memorandums are usually strictly business, highly technical, and concise; but they must, of course, be adapted to the intended reader.*

It has been said that the various levels of government and all major businesses in the United States would come to a screeching halt if their employees quit writing memorandums. Memos are the lifeblood of a large organization. They transmit information from worker to supervisor and on up and down the organizational structure, from one department to another, and from one branch to another. Many records that are kept on microfilm or in a computer data bank were originally written as memorandums.

Purpose of the Memorandum

Memorandums are often called interoffice letters. They are that and much more. Here are some of the things memorandums do:

1.   *Act as letters.* Use a memo for someone inside your organization in the same way you would use a letter for someone outside the organization. (Some organizations do use letters internally for very formal situations, however; and memos may sometimes come from the ouside.)

2.   *Furnish a record of work accomplished.* When you want to remember exactly what was done, write a memorandum to summarize the work. If several people are involved in the project, send each a memo. You can even write a summary strictly for your own files.

3.   *Keep complex information straight.* Complex, highly technical material is often difficult to remember precisely if it is not written down. Again, use a summary memo.

4.   *Disseminate information to large numbers of people.* You can duplicate a memorandum to communicate with a large group.

5.   *Record oral discussions.* You can easily waste time and paper trying to get everything in writing, but when you discuss important subjects or make decisions on the telephone or in group discussions, you should keep a written record. Today you may be convinced that you will remember exactly what you and a colleague are discussing, but next year or next month your memory may not be so clear, and your colleague's memory may not agree with yours. Send him or her a summary memorandum and keep a copy for your files. In group conferences, one member can be designated to send the others a summary memorandum.

6.   *Submit reports.* Memos are commonly used for reports that do not require formal presentation (see Chapter 10) and are not done on preprinted forms. Memo reports generally cover minor projects or studies or

phases or segments of longer projects. Memorandum reports are discussed in a later section.

Memorandums are often useful and necessary, but too many are written. Many managerial and supervisory employees are kept so busy writing and reading memos that they have little time for other aspects of their jobs. Recently, for instance, a midlevel engineering supervisor at a small governmental installation admitted to writing more than five hundred memos and receiving more than a thousand last year. No wonder he complains of being overworked!

Write a memo only when you have a definite reason for doing so, not because it seems like a good idea or because someone else would probably write one. If the intended reader has no need to receive the information and you have no need to give it, spend your time on something more useful. If your colleagues send out reams of memorandums, let them. You restrain yourself. The restraint will be appreciated.

Not only are many memorandums written unnecessarily, but many are sent unnecessarily. Many technical people mistakenly send copies of their memorandums to nearly everyone. Company policy or procedures frequently require employees to send copies of certain memorandums to supervisors and various high-level officials. That you cannot avoid, but you can limit copies to those people specified by company policy and those directly involved in the content of the memo.

Guard against the tendency to send unnecessary memos in an attempt to seem busy; don't equate quantity of paperwork with quality of all work. Write memorandums wisely; use them when they fulfill a legitimate purpose, and send them to readers who need them; don't fill your colleagues' waste baskets with them.

Memorandum Form

In place of the dateline, inside address, salutation, complimentary close, and signature commonly used in letters, a memorandum requires only a simple heading. Most organizations furnish special memorandum paper with a heading already printed. You simply fill in the required information behind each item in the heading, then proceed to develop the body. If no such printed sheets are furnished you can easily type or write your own. Various terminology is used for the heading entries; select your own, making sure you allow for the following information:

1. date written
2. names of primary readers
3. names of people getting copies
4. your name
5. the subject of the memorandum.

INTEROFFICE COMMUNICATION

BRIARFIELD POLICE DEPARTMENT

Highway 441   Briarfield, Florida 32791

888-5000

To: **Maurice Berg, Police Chief**
By: **Perry Hinden, Sgt.**
Date: **13 August 1980**
Re: **Summary of August 12 discussion on public relations**

After discussing several plans for improving relations with the community, we reached the following conclusions:

1. Two age groups, 18–25 and 11–15, need our primary attention. Bill Wood's health education class is doing wonders with the 16 and 17-year-olds.

2. We must stress our positive accomplishments, show how we cleaned up Oak Hill and stopped the troubles at the Crazy Eight.

3. Tracey Patterson and Mike Perinelli should be approached about giving lectures. Faye Zellman and Paul Holler simply cannot relate to juveniles and minority groups.

4. A specific plan must be ready for submission to the city council by September 1.

Figure 9.1 Sample Memorandum 1

Some typical headings are found on the sample memorandums in this chapter. (See, for example, Figures 9.1–9.3.)

The form of the body of a memorandum is similar to that of the body of a technical letter. Ordinary paragraphs are used much the same as in other common written forms. Paragraphs are normally full-blocked (no indentation) but may be indented.

If you wish you may sign your memorandum in the same general area where you would sign a letter. Do not use a typed signature or complimentary close. Some people prefer not to sign memorandums, instead merely writing their initials after their names in the heading. Many others neither sign nor initial memorandums. Use the method you prefer or follow general practice in your organization.

Unless a memorandum is extremely informal and is being sent to someone you know well, it should be typed. Try to type your memorandum as well as you can, and make it attractive. If the memorandum is going to an important official or a large group or if it is going to become a permanent record, you should take the same pains with it as with any important document. Otherwise,

```
                      *  MEMORANDUM  *
                      QUIX STOP FOODS
──────────────────────────────────────────────────────────
    TO:     David L. Lacharite         FROM: Absalom Wills
    SUBJ:   Change in Operating Hours  DATE: September 6, 1981
    COPIES: Distribution List


    Effective Monday, September 30, we will close at 10:00 p.m. instead
    of midnight.  We will continue to open at 6:00 a.m.

    Our recent audit indicates that we have not been making expenses
    after ten o'clock, and we have no reason to believe that situation
    will change.  Please be certain that all department heads and super-
    visors on the distribution list are made aware of this fact so that
    they can explain to affected employees.

    The following personnel scheduling changes should be made:

        1.  5:30-2:00--no change
        2.  8:00-4:30 -- no change
        3.  4:00-12:30-- 3:30-10:30
        4.  8:00-12:00-- discontinued: transfer employees to another
                                       shift if possible

    Notify all affected employees next week so that those leaving will
    be given adequate notice.
```

Figure 9.2 Sample Memorandum 2

make it neat and attractive but do not waste time retyping to try to achieve perfection. In fact, many organizations will not allow their employees to retype or to make costly corrections in ordinary memorandums.

Memorandum Style

No perfectly standard memorandum style exists; the appropriate approach in one situation for one reader may be totally inappropriate in another situation with another reader. Any time you write a memorandum, you should ask the same preliminary questions and do the same preliminary planning as you would in preparing anything else. Take the time to plan all memorandums, even the simplest and most common. If a document goes out to someone with your name on it, you should want it to be as effective as possible. Preliminary planning will help ensure that effectiveness. Check Chapters 1 and 2 and use a planning worksheet.

Despite the unique nature of every memorandum, some general guidelines should be followed.

INTEROFFICE MEMORANDUM

● JANITROL
● FEDDERS
● GIBBSON

C & C Heating and Cooling

143 West Wolverine Rd.
Kenosha, Wisconsin 50336   *884-7890*

To: Bob Tucker, Sheet Metal Shop Foreman

From: Barbara Dexter, Managing Partner

Subject: Request for recommendations on planning for new shop

Date: September 17, 1981

Please send me a complete recommended design and equipment
list for the new shop in our Lakeview branch by Tuesday,
October 3. This is obviously short notice, but we have to move
fast. The zoning board gave us the go-ahead and Tompkins
Construction is ready to begin.

As we have discussed before, specify the best equipment you
know of, and plan for anticipated growth. We want a first-
class shop.

I should be able to give you approval to order equipment by
October 15.

Figure 9.3 Sample Memorandum 3

1. Memorandums are generally strictly business. Goodwill endings that are appropriate in letters are seldom used in memorandums. Avoid making small talk or sounding chatty unless you are certain that is what your reader wants.

2. Conciseness is an even greater virtue in memorandums than in most other written forms. Try to determine the essential information and omit everything else.

3. Use the level of technicality that is appropriate for your intended reader. More memorandums are highly technical than nearly any other form of writing.

4. Keep in mind that you will often be writing a memorandum as much for your own future reference as for the designated readers' use. Construct it so that you can understand it weeks or even years in the future.

Memorandum Reports

As mentioned earlier, memorandums are used for many technical reports. Long reports (discussed in Chapter 10) are often done up formally, with separate title pages, tables of contents, and other special parts. Shorter reports are usually submitted on printed forms or as memorandums.

A survey of all the different types of memorandum reports written in various fields would turn up hundreds, probably even thousands, of different kinds— one common type, in fact, would be a survey report documenting the results of such a survey. Here is a brief sampling of some other common types:

accident report	equipment utilization report
audit report	feasibility report
breakdown report	travel report
hourly report	investigative report
daily report	inventory report
weekly report	case load report
annual report	test report
incident report	inspection report
staff utilization report	appraisal report
evaluation report	study report

No one could become expert in writing every type of memorandum report. But you can master those written most commonly in your profession by analyzing the reader and the writing situation and planning carefully.

Basic Memo-Report Pattern

1. Heading ● Many memorandum reports can be submitted on printed memo pads, using standard memorandum headings such as *To, From, Date,* and *Subject.* The subject entry then becomes the title for the report. (See the first model below.) Other times you will want to construct your own headings to give more thorough identifying information. Note, for instance, the second model, which uses the entries *Area inspected, Date of inspection,* and *Inspected by* in addition to variations of the standard entries. The important point is this: Use a heading that will tell the reader at a glance what the document is and who is sending it.

```
     To:  Roscoe Gereau              Date:  May 19, 1980
   From:  Al Harrison
Subject:  Preliminary report on feasibility of sewer-line
          extensions
```

```
Area inspected:  A.V. support equipment
Date of inspection:  10-3-80
Date of report:  10-7-80
Inspected by:  Carroll Miller, J. C. Merriweather, Anita Snyder
Reported by:  Anita Snyder, Senior Technician
Copies to:  Dan Ricks, Bldg. Superintendent
            Jennifer Dahlberg, Dir. of Institutional Resources
```

2. Introductory Statement ● Tell the reader briefly why you are submitting the report and exactly what you are reporting on. A detailed introduction giving complete background information is necessary only in a long, formal report. Readers of the short memo report want a quick overview of its intended function. They usually know the background situation; if not, they can ask questions later. One, two, or three carefully phrased sentences should suffice. Here are introductions for the two model headings shown above:

<u>Purpose</u>

This report summarizes my findings after one month's investigation of the feasibility of expanding the Oreana sewer system into Greenswitch and Plainville. A final report and a firm recommendation will be submitted in approximately one month.

<u>Introduction</u>

Mr. Merriweather, Ms. Miller, and I conducted a complete inspection last week (10-1, 2, 3) of our audio-visual facilities. This report summarizes our findings and offers several recommendations for repair and replacement.

---

3. Findings or Results ● The longest and most important section of a memo report is the presentation of findings or results. Their number and nature will vary considerably according to the type of project or study; but in nearly all instances, your findings or results will be the heart of your presentation. Select the clearest, most concise method of presentation you can. Try using a table if you have many specific facts or figures, or a list if you cannot use a table. Do not waste the reader's time by burying your raw data in elaborately constructed paragraphs. Here are findings and results from the two models:

4. Conclusions and Recommendations ● After you present your findings, you are in the best position to draw conclusions from your information. Although you are not expected to analyze your data so thoroughly as to draw every conclusion possible, you should present what you consider the most significant implications of your findings.

The next step is to offer recommendations based on your conclusions. As the person who knows your findings best, you are the logical person to propose such specific recommendations as the kind of further action to take or to avoid.

This section of your report must be based on sound reasoning. Even if your facts are accurate, logical reasoning is necessary to translate those findings into sound conclusions and recommendations. A review of the discussion of logic in

Findings

	Plainville	Greenswitch
Mileage from plant	3.1	7.6
Number of initial hookups	56	82
Possible total hookups (next ten years)	91	118
Initial cost	750K	1440K
Revenue from hookup fees @ $4000	224K	328K
Revenue sharing	250K	450K
Required indebtedness	276K	662K

Results

Machine	Condition
B-H 186 Super-8	New
B-H 185 Super-8	Excellent--needs cleaning
B-H 184 16	Needs factory overhaul
B-H 183 16	Needs factory overhaul
B-H 182 16	Manual threading, good condition
P.S. 180 Cassette	New
P.S. 179 Cassette	New
P.S. 178 Cassette	V-good, needs cleaning
S 177 Cassette	Good
S 176 Cassette	Needs minor repair
K 175 35 with Car.	Excellent
K 174 35 with Car.	Very good
K 173 35 Manual	Good
X 169 Overh.	Good
X 168 Overh.	Good
X 167 Overh.	Needs minor repair
X 166 Overh.	Beyond repair
RCA 147 VTR	New
RCA 146 VTP	New
RCA 145 VTM	New
RCA 144 VTEd.	New

Chapter 2 can be helpful here. Next are sample conclusion and recommendation sections. The first sample is from a preliminary report; thus, it offers only a few tentative conclusions and no recommendations. More fully developed conclusions as well as more detailed recommendations would be included in the final report.

Conclusions

1. Plainville looks promising. If our bonding power is sufficient
   we should strongly consider expanding there.
2. Greenswitch does not look so promising. Certainly, 662K is more
   than we can long-term bond; thus the rates would have to be
   exorbitant. I'll continue my investigation, but it does not
   look promising.

Conclusions

1. We are well fixed for movie units if we can get 183 and 184
   overhauled.
2. We are also well fixed for audio-recorders.
3. We are well fixed for slide projectors. All three are first-
   rate.
4. With only two usable overheads and one in need of minor
   repair, we have a definite lack.
5. The one usable opaque is sufficient.
6. The video unit is, of course, more adequate

Recommendations

1. Have all equipment that needs service repaired or overhauled
   immediately.
2. Discard all equipment indicated as beyond repair.
3. Purchase two new overheads immediately.
4. Investigate service contracts to keep our equipment in better
   condition.

Approach and Style

The memo report is one of the easiest forms of technical writing to compose. It is usually written for one reader, most often someone within your organization. Designing a report for such a reader will take some work, but it is certainly less difficult than writing for large groups or outsiders.

The term *informal report* as a synonym for memo report indicates the basic approach of this form of communication. Use straightforward language appropriate to your reader's level of technicality and short, simple sentences arranged in a number of brief sections, each covering a separate aspect of the subject.

Include only information essential to your purpose; leave out information that is not clearly relevant. (Of course, you should not go to the extreme of omitting important information.) Goodwill introductions or conclusions are not necessary. Say only as much as you have to in order for your reader to grasp your major points.

Informal implies that standards of mechanics, usage, and form are more relaxed than for a formal document, but it does not mean that you can disregard rules and conventions. Check your spelling, punctuation, and grammar. Make your report neat and attractive. The reader may not get upset if your copy has one or two minor errors, but why gamble? You are not going to get into trouble for being too precise or too correct. Most informal reports are typed, but frequently a handwritten one is permitted. When you are in doubt, type your report. Figures 9.4 and 9.5 illustrate the form and style of memo reports.

INTER-COMPANY COMMUNIQUÉ

TO: Rence LaBeque, Managing Partner
DATE: 11-14-81
FROM: Sam Boureajis
RE: Recommendations on Wine List
for Chateau Richebourg

Purpose of Report: To present specific suggestions for a wine list at our new Marietta outlet--Chateau Richebourg. The report is based on a three-month study in Marietta, major French restaurants in New Orleans and New York, and the Bordeaux and Burgundy regions of France.

Findings:

1. Existing Marietta area restaurants either have very limited wine lists or stock primarily domestic wines.

2. The better Marietta establishments do have frequent calls for expensive French wines.

3. Restaurants in New Orleans and New York are receiving many calls for moderately priced French wines. Diners who formerly drank domestic wine or none at all are becoming more aware of French wines. Diners who formerly asked only for La Tour, Lafite, Margaux, etc. are now asking for good, reasonably priced brands. At the nine places visited, total wine sales are up forty-seven percent over two years ago.

4. B&C and B&G can ship us all of the wines we can use.

Conclusions:

1. We must stock a limited amount of vintage year first-growth Bordeaux and Grand Cru Burgundy. Dom Perignon and similar champagnes must also be kept available.

2. A large supply of moderately priced wine is needed. Our clientele will need some education, but Louis Vigency and his sommeliers can handle that.

3. Our previous notion of keeping minimal domestic wines is sound. Rob Short will give us a small list of the finest California and New York wines for customers who ask for domestic.

Figure 9.4 Model Memo Report

page 2

Recommendations:

    1. Add anything you want to the attached order forms, then send them out immediately. Delivery will require sixty days.

    2. Have Vigency send me updated inventories monthly so that I can adjust our stocking patterns as needed. The supply indicated on the attached order blanks should last six months, but I want to order monthly to keep a close rapport with our suppliers.

    3. The following is a list of wines Vigency should push. Have him go over it with his entire staff, especially for coordination with Lapiererne's menu.

Red Bordeaux

Bordeaux Superieur, Beaulieu
Bordeaux Superieur, Bichot
St. Emilion, Bichot
Chateau Corbin Vieille Tour
Chateau Laffitte Cantegric
Chateau Moulin De Marc Graves
Chateau Kirwan Margaux
St. Julien, Bichot
Medoc, Dreyfus Ashby
Chateau Savoie

Red Burgundy

Pommard, Bichot
Beaujolais-Villages, Drouhin
Cote de Beaune-Villages, Drouhin
Clos De Vougeot, Faively
Mercurey "Clos de Myglands"
Gevrey-Chambertain, Faively
Moulin-a-Vent, Bichot
Vasne Romanee, Bichot
Cote de Nuits Villages, Bichot
Richebourg, Domaine du Clos Frantin

Rosé

Tavel Delas Frères
Rosé d'Anjou, Ackerman-Laurance
Macon Rosé, Bichot

Sparkling Wines

Sparkling Burgundy, Bichot
Krug Champagne
Dom Perignon

White Wines

Petite Chablis, Bichot
Pouilly-Fuisse, Drouhin
Pinot Chardonnay, Bichot
Soleil Blanc, Drouhin
Chablis Grand Cru-Moutonne
Meursault, Bichot
Laforet Macon-Villages
Chablis Grand Cru-Les Vaudesirs
Sauternes, Dreyfus Ashby
Barsac, Bichot
Sauvignon Blanc, Bichot
Graves, Dreyfus Ashby
Graves Superieres Sec

Figure 9.4 (Continued)

Inter-office Memorandum   **CRAMER—MICKEL ARCHITECTS**

To: Vernon P. Sylvia, General Manager
Date: 10/10/81
From: Michele McDonough
Subject: Dallas trip to study Tudor designs

Introduction

On Monday morning, October 7, I went to Kelley Bros. in Dallas,
Texas, to study their innovations in Tudor designs. I returned
Wednesday morning, October 9.

Findings

James Gore and I spent twelve hours Monday and Tuesday in consul-
tation. He explained step-by-step how they drew up the Sir Ger's
Steakhouse plan in Atlanta. Complete working and finished drawings
are attached.

Connie Driggers of Kelley Bros. legal division signed a release
and accepted our check. The release has been given to Marv Pruitt.
A copy is attached.

Conclusions

1. Bonnie Martin, Ralph Chittendon, and Jerry Hartje can handle
    the new design with about an hour's instruction.

2. No major changes in the "Sir Ger" plan are needed to meet our
    needs.

Recommendations

Let's move now. I can get the group working on it Monday morning.

Enclosures: Copy of Kelley Bros. release
             Copies of working and finished drawings #12B218
             Form 316A, Expense voucher

Figure 9.5 Model Memo Report

A Portfolio of Memorandums

Josephes Industries
Memorandum

To: All exempt employees   From: Marisa Jones, Benefit Office

Date: November 1, 1981   Subject: New dental insurance
                                             benefits

Congratulations on your excellent insurance record. Because of
our good health and restraint, we will soon be given 80 percent
dental insurance for the same price we are now paying for 50 percent
coverage.

The increased benefits will go into effect on January 1, 1982. Our
carrier, Wilmington National Life, will be distributing new insurance
booklets and claim forms next week.

Call my office, extension 456, if you have any questions.

Figure 9.6

Leaf Lovers' Nurseries, Inc.
Davenport, Bettendorf, Rock Island, Moline
Intercompany Memorandum

To: Greenhouse managers   Copies to: J. Cenko, Vice-President
                                               P. Cappelo, Chief Grower

From: L. J. Susskind,   Date: April 21, 1980
       Operations Officer

Subject: Reporting procedures for controlled pesticides

    Attached is a new form to be used in recording our use of pesti-
cides. As you know, the EPA and state conservation departments
continue to change their regulations on the use of pesticides. Because
keeping track of these changes is becoming increasingly difficult, we
have decided to monitor our use even more closely. Note that the new
form asks you to record the following information:

    1. Gallons of each substance on hand at beginning of month.
    2. Date and amount of each application.
    3. Gallons received.
    4. Gallons on hand at end of month.

    Keeping this record is extra work for all of us, but we really
have no choice. If you have any suggestions for improving the form
or the reporting procedures, let Ms. Cappelo or me know. Otherwise,
we will expect your May report by June 7.

Figure 9.7

J & D Flight Service

17 Herndon Airport
Fort Smith, Arkansas 72830

To: All employees   Date: July 17, 1980

From: J. B. Exum, Managing Partner

Subject: August 12 visit by Beechcraft representatives

    Joe Plochman of Beechcraft will be here all day Thursday,
August 12, to demonstrate three new models we are considering for
possible purchase. Please arrange your schedules so that you can
spend some time with Joe and his staff, examine the aircraft, and,
if appropriate, try them out.

    I'll be talking with some of you personally in the next few
days, but I value everyone's input. These planes might be just
what we need, and the more information we have the better able we
will be to decide for sure.

    Feel free to adjust your schedules so that you can help us in
making this big step forward for J & D Flight Service.

Figure 9.8

Tarrant County Sheriff's Department

Fort Worth, Texas 79633
Interdepartmental Memorandum

To: Capt. Stanley Pascevich   Date: 7-17-80

From: Lt. JoAnn Hardee *JH*

Subject Planned efficiency measures in CID

This is the summary memorandum you requested of yesterday's planning session. It should furnish a good basis for your discussion
with Sheriff Perry and Major Thomason.

Increased use of unsworn personnel

1. To conduct telephone inquiries to pawn shops
2. To schedule interviews with plaintiffs, witnesses
3. To complete routine reports, exact reports involved to be
    determined as soon as possible

Changed hours of detectives

1. Start changed from 0800 to 0900
2. Stop changed from 1700 to 1800

Procedural changes

1. All sergeants will have full scene jurisdiction unless homicide,
    rape, or juvenile felonies are involved.
2. Rape United (headed by Sgt. O'Banion) will report directly
    to Lt. Hardee.

Anticipated results

    While not making up for the additional staff we desperately
need, the planned changes should help us to reduce current caseload
by almost 20 percent. Other proposed changes did not stand up under
scrutiny and will not be implemented.

                                            J. A. Hardee
                                            7-17-80

cc: Sheriff Maurice Perry
     Major Lonnie Thomason

Figure 9.9

FALCO ELECTRONICS

A Division of Tinkertown Enterprises

Interoffice Communication
To: All employees   Date: 9-23-80

From: Jim Bramer, Assistant General Manager *JB*.

Subject: Temporary Changes in Payroll Deductions

Tinkertown has agreed to let us keep our unique system of unlimited payroll deductions. Effective January 1, we can resume our
former deductions or expand them if we wish.

For the remainder of 1980, we will, unfortunately, have to get
along without any deductions **except FICA, retirement, and federal
with**-holding tax. This will give Tinkertown time to develop the
necessary software for their Houston computers.

Thanks for bearing with us. You've all been great in making the
transition.

Figure 9.10

Likins, Massey, Yetter Associates
MEMORANDUM

To: Brian Ponewash, Interiors   Date: March 19, 1980

From: Bella Yetter, Operations

Subject: Your proposed changes in Wilderness III model

     Your proposed changes in interior decor for the Wilderness
III model in Oak Harbor show exactly the kind of creative
thinking we hired you for. I've shown them to Mr. Likins and
he agrees.

     However, our market research as well as our past experiences
in this particular market dictate that we stay with the finished
oak and new brick. Most buyers in this market own Early American
or Traditional furniture and demand the more polished look.

     Don't let this get you down. We do hope to use your ideas
in Baytree where the market will be a bit more affluent and
willing to experiment.

     Keep up the good work.

Figure 9.11

Florida Cedars Hospital
Bithlo, Florida 32711

Department of Patient Education   MEMORANDUM

To: Lyle Merino, Instructor Trainee   Date: January 27, 1981

From: Betty Meitin, Parent Education Instructor *BM)*

cc: Lois Hall, Director of Parent Education

Subject: Instructions for Wednesday night Lamaze class, February 4

    Thanks for agreeing to take over on such short notice. Having
seen you work so effectively with Mrs. Hall's group, I know you will
get along just fine. Since my schedule is slightly different from
Mrs. Hall's, I've listed the activities that you'll need to cover.

1. Show "The Birth of Eric."
2. Give a ten or fifteen minute break.
3. Review orally all four breathing techniques.
4. Practice relaxation--three minutes.
5. Practice deep-chest--three minutes.
6. Practice shallow-chest--three minutes.
7. Demonstrate pyramid, then practice--fifteen minutes total.

    Several class members will need help with their effleurage but
all are conscientious and cooperative.

    Call me before Friday morning if you have any questions--my
plane leaves early. Mrs. Hall will be glad to answer any other
questions.

    Thanks again.

Figure 9.12

Costly and complex problems in a highly technical project, such as the installation of a sound system, may be avoided if a thorough and precise memorandum has been prepared.

• CASE STUDY

Assume that you are a sound engineer supervising installation and repair work for a new, ambitious company. You have recently installed a complex, expensive sound system in a new disco club, Studio Z. Within three weeks of its opening, the club is already having major problems with the system. Your company's managing partner, Brice Sutter, has demanded that you drop everything else and immediately get the system working properly. Bad publicity about the Studio Z job would seriously threaten your company's chances of becoming well established in the community.

Write Sutter a memorandum explaining that you are certain the problem is a result of the mismatched components that he and the club's owners, Cheryl Sledge and Donald Summer, insisted on using. Point out that it will be necessary to change components and work your technicians overtime if he wants the system repaired promptly.

177

Next, assume that you have completed the job. Write Sutter a memorandum report (making up any figures you need). Recommend as diplomatically as possible that he consult you about compatibility before selling any new systems.

● EXERCISES

1. Write a summary memo recording the work you have done so far in this course. The memo would go into your files for future reference.

2. Write a memo to your instructor summarizing your work thus far in the course.

3. Write a memorandum summarizing the next meeting or conference you attend. Any group meeting will do: club, church group, department at work, even class discussion.

4. Write a memorandum to your instructor suggesting a topic for a major writing or speaking project you will be doing later in the term.

5. Write a memorandum to your supervisor at work. If you are not currently employed, assume you have a job related to your major field. Politely suggest at least two changes in the way your department is operated.

6. Write a memorandum reporting on a recent hypothetical—or real—business trip. Use Figure 9.5 as a model, but adapt it to your type of trip.

7. Write a memorandum of instruction showing a new employee how to do one of the tasks comprising your job.

8. Write a cover memorandum to accompany the other memos you are sending to your instructor.

10

Long Reports

Overview   *Long, formal reports may include extensive front matter—title page, contents, approvals, distribution list, preface, list of illustrations, credits, abstract or summary—and back matter—reference list, appendixes, glossary, and index—in addition to the main text. The term* semiformal reports *describes a broad range of reports lying between the memorandum report and the long, formal report. Progress reports, which document a part of a study or project, are flexible as to length and degree of formality but have certain conventions of their own.*

The short, informal memorandum report seldom contains more than three or four pages; the long, formal report can have as many as several hundred; and between these two extremes are many possible combinations of length and formality. *Semiformal reports* run the gamut from slightly "prettied up" memorandums to borderline formal reports. *Progress reports* can be short and informal or long and formal, but they have certain conventions of their own.

Long, Formal Reports

While short reports submitted as memorandums or on printed forms are frequently required of nearly all technical and professional people, the same is not true of long, formal reports. Some people have to write them regularly, some occasionally, and some never. Jobs involving research or a great deal of supervised work and those done contractually for other organizations most often require long, formal reports.

The greater length of such reports is necessary for two reasons: they generally cover major jobs, projects, or studies that cannot be adequately explained in short reports; and they contain types of material not included in short reports.

Nearly every large organization has strict guidelines for the format of long, formal reports written by its employees or contractors. Always check such guidelines when you begin your preliminary planning. If none are available, check with the appropriate professional organization. Many groups, such as the American Chemical Society or the American Management Society, offer useful guidelines for setting up formats. If left on your own, use the suggestions here, look at reports done by your colleagues, and check with your supervisor.

A long, formal report may contain as many as twenty-five parts, not including sections and subsections. These parts can be classified into three categories: front matter, main text, and back matter. In addition, a letter of transmittal should accompany the report.

Letter of Transmittal

The letter of transmittal is sent with the document to introduce it to the reader. Although it is discussed here, it is not really a part of the document. It accompanies the report, attached to the outside or atop it in a large manila envelope. A letter of transmittal, like most other types of letters, contains a dateline, inside address, salutation, and other conventional parts.

The body of a well-organized letter of transmittal has three sections: an introductory paragraph which identifies the document being transmitted, a paragraph or two calling one or two significant features to the reader's attention, and a concluding paragraph which expresses an offer to answer questions or discuss the matter further.

The most important and difficult section to compose is the second one. This section is challenging because the features to be mentioned must be selected with the reader clearly in mind. Emphasize the features that a reader will find important or interesting. If the report is going to several readers, then you must compose a separate letter of transmittal for each. If you have a large number of readers, classify them into a few groups and write one version for each group.

Selecting the best details is not enough; you must express each one in the best way. Give the page or section number of each item mentioned, providing just enough information so that the reader will turn to that section or page.

Take your letter of transmittal seriously. A good one can entice readers into looking at the report immediately and can increase the chances of their viewing the entire document in a positive light. A perfunctory letter of transmittal has the opposite effect. Here are the main sections from two letters of transmittal that introduce the same report to two different readers:

```
Dear Mrs. Garonski:

     The accompanying report contains the results of our recent
video player comparisons.  Recommendations for acquisition are
included as you requested.

     Section 5.3, which contains complete cost analyses of every
player on the market, should be especially useful to you in
justifying to Colonel Blake our new budgetary requests.

     Notice also the table on page 33, which compares tape versus
super-eight performance.  It documents our belief that super-eight
would perform just as well.

     Please let me know if you need any further data or if you
wish any changes or corrections made.

                              Sincerely,
```

```
Dear Colonel Blake:

     Enclosed is Report #371B, "Performance Characteristics of Video
Players," documenting tests recently completed by our division.

     You might find Section 5.3 helpful in planning the upcoming
budget, as it cost analyzes every player currently available.  Also
important is Section 3.1, which describes salient features of each
player.  This should be useful in planning our new public relations
campaign.

     Please call on me if you have any questions or suggestions.
I will be very happy to work with you further on this matter,
either individually or through our civilian liaison Betty Garonski.

                              Sincerely,
```

KIDDIE COLLEGE

small fry

NADINE ANDREWS, DIR

4116 MAPLEWOOD LANE

LINCOLN, NEBRASKA 73716   323-733-3333

September 9, 1981

United States Department of
   Health, Education, and Welfare
Department 603-16
2400 Pennsylvania Avenue
Washington, D.C. 00017

Gentlemen or Ladies:

    Enclosed is the report you requested as part of our special
services contract #816A113. The report conforms with all
stipulations outlined in your March 17 directive to child-care
contractors.

    You should find Section 3.2, pages 11 through 14, especially
useful because it gives a comprehensive inventory of our facilities,
complete with color photographs. Page 19 should also interest you,
especially the table of family income.

    I hope you are pleased with the report and with our program.
Should you need any further data, let me know.

                          Sincerely yours,

                          Nadine Andrews

                          Nadine Andrews
                          Director

NA:bl

Enclosure

Figure 10.1 Model Transmittal Letter A

KIDDIE COLLEGE

smɑll fry

NADINE ANDREWS, DIR

4116 MAPLEWOOD LANE
LINCOLN, NEBRASKA 73716   323-733-3333

September 9, 1981

State Department of Sanitation
State Office Building
142 South Court Street
Lincoln, Nebraska 73703

Gentlemen or Ladies:

Enclosed is the copy of our annual report to the Department of Health, Education, and Welfare that you requested.

In addition to noting throughout that we meet all federal, state, county, and local regulations, you might find two sections particularly useful: Section 3.4 shows our kitchen facilities to be among the finest available, and Section 3.5 shows our restrooms to be far above the norm in both cleanliness and size.

If you would like to discuss any aspect of our operation, call me at 323-316-3131 or drop by our center at any time.

Sincerely yours,

Nadine Andrews

Nadine Andrews
Director

NA:bl

Enclosure

Figure 10.2 Model Transmittal Letter B

Front Matter

Front matter is everything that precedes the first page of the main text. A long, formal report will require at least three, and perhaps as many as twelve, pages of front matter.

Here is a list of the essential, as well as the optional, parts of the front matter.

1. Cover Sheet ● Many organizations require a special sheet attached to the outside front cover. This cover sheet contains the same basic information as the title page. If your organization wants a cover sheet used, it will either supply one with blanks to be filled in, furnish a list of items to include, or provide actual models. But do not bother with a cover sheet if you do not have to.

2. Title Page ● The first page of a formal document is nearly always the title page, which is essential both as a matter of convention and as a guide to the reader. Many organizations use carefully designed title pages with blanks that the writer merely fills in. If no standard page is available, design your own, keeping two purposes in mind: completeness and neatness. The following items are usually included:

a. Title of report.
b. Author's name.
c. Date submitted.
d. Company name (and often address).
e. Department or division name.
f. File number assigned the report.
g. Name of person authorizing the report.
h. Space for signatures of those who must approve the report.
i. Name of the primary reader and anyone else being given copies.

A typical title page is included in the model report at the end of this chapter. Adapt it to your own situation, adding or deleting information as you see fit. If you design your own, be sure to include all the relevant data, and arrange it on the page to look neat and attractive.

3. Approvals ● It is sometimes necessary to have several people approve a report before it can be officially submitted. Some organizations want a separate page set aside for the signatures and possible comments from those who approve the report. Do not include such a page unless you are required to.

4. Distribution ● Some reports are intended for so many readers that their names cannot be conveniently included on the title page. When this happens, use a distribution page. List the primary reader on the title page and the others on the distribution page. Be careful, though, not to offend someone inadvertently by placing his or her name low on the list. It is better to list several primary readers than to offend high-ranking people by listing them under distribution or carbon copies.

The distribution page should be entitled "Distribution" and should contain only the list of names. They can be arranged by rank or by alphabetical order as you see fit.

5. Preface ● A difficult problem in developing the front matter is distinguishing between a preface and an introduction. A preface is part of the front matter; it contains explanatory material about the document itself. An introduction is the first section of the main text; it usually provides background information on the report. Use a preface to discuss, for example, important distinctions between this and similar reports, unique formats, levels of technicality, or anything else the reader would find helpful before he gets into the main text. Even though it performs a useful function, a preface is only occasionally used.

6. Table of Contents ● The table of contents of a report shows what the document contains. In constructing the table of contents, list in order the sections and subsections into which the main text and the back matter are divided, and add the corresponding page numbers. The sample long report in this chapter contains a typical table of contents.

7. List of Illustrations ● In many reports the table of contents is followed immediately by a table of illustrations. To construct one, list each illustration in the order in which it appears in the main text and give its figure or table number, title, and page number.

8. Credits or Acknowledgments ● When individuals or organizations furnish large amounts of material, advice, or facilities for your report, you may want to acknowledge their assistance formally. This procedure is sometimes required by law if your report is to be printed or copyrighted. In a shorter, less formal report, you can include acknowledgments in your preface or even in your introduction.

9. Abstract or Summary ● The abstract or summary is probably the most important part of a formal report. After glancing at the title page and perhaps checking through the table of contents, the typical reader will turn to the abstract or summary. Unfortunately, many readers will go no further. In that case, it is imperative that they find your most important ideas in the abstract or summary. Further, if you write it well you can sometimes impress readers sufficiently that they will read other sections of the report.

While both abstracts and summaries are drastically shortened versions of the main text, this book makes an important distinction between them. The abstract is a preview or *overview* of the main text that follows. It states the kind of information and conclusions the main text covers but does not present any of them. However, your most important conclusions *should* be presented in the summary. A summary says just what the main text says, in a greatly condensed form. Remember, though, that some people use *abstract* as it is used here, to designate a brief description of the main text; others use *abstract* and *summary* interchangeably; and still others use *descriptive abstract* for our *abstract* and *informational abstract* for our *summary.* If you are asked to write a summary or an abstract, make sure you find out exactly what the term means in your organization.

The suggestions and models below should help you learn to develop effective abstracts and summaries. If you can follow these suggestions and if you are not afraid to ask a few questions, you should be able to handle any situation that arises.

Abstract of Chapter 10

This chapter begins by defining and analyzing a formal report, then showing how to construct each part, emphasizing abstracts and summaries. Then it explains and exemplifies the construction of semiformal reports and progress reports.

Notice that the sample abstract is only two sentences—thirty-four words— long. This is typical. Rarely do abstracts contain more than a hundred words. They are short, but their apparent simplicity is deceptive. At first glance, the abstract may seem to involve merely scribbling out a few sentences to describe the contents; but a closer examination will show that each sentence has been carefully constructed.

Notice particularly the choice of words. *Analyzing,* for instance, tells the reader that the formal report is taken apart and examined closely. The more commonly used *covering* or *discussing* would have been less effective, because neither would have given the reader as much help in determining what the chapter was about. A good abstract not only identifies the subjects treated but tells how they are treated. For example, instead of telling your readers you have covered certain methods of production, tell them you have compared them, illustrated them, or even just listed them. Concrete words such as these tell readers at a glance whether the report is likely to include what they are looking for, thus helping them to decide whether to read it.

To develop an effective abstract for your report, begin by checking your outline or table of contents for accuracy—you can use either or both to develop the abstract. Then compose a "capsule," or main sentence, that identifies the overall purpose of the report. After this, write sentences describing the function of each major section. Now see if some of the sentences can be combined. You might find, for instance, that sections 3.0, 4.0, and 5.0 deal in the same manner with class 1, 2, and 3 fires. One sentence could easily describe all three sections. This technique enables you to abstract even very long reports into a few sentences.

Now check each sentence for conciseness and concreteness. Make each sentence as clear and crisp as possible, but do not omit words just to be brief. The abstract should be concise, but it must be clear and grammatical. And remember to use concrete verbs.

Developing an effective summary begins in the same manner. From an accurate outline of the main text, develop sentences for each point. Do not describe each section; state its main idea. Do not talk about the report at all; rewrite it, using one-twentieth to one-tenth the number of words. Your completed summary should read like an independent composition that expresses the same basic ideas as the main text of the report. Take a look, for instance, at the chapter summary on page 180. Notice that it makes no mention at all of this chapter.

A well-organized text makes developing a summary much easier. You can often pick main ideas or topic sentences directly from the text. If you cannot find such sentences in yours, you will not only need to develop some for your summary, but you should also consider inserting them in the text. In fact, many good writers use the development of summaries as checks on the organizational effectiveness of their texts.

Edit your draft twice. First, check your coverage. Have you covered all major points? If not, make the necessary additions. Have you included unnecessary points? Developmental details rather than main ideas? If so, make the necessary deletions. Now try to combine sentences so that one good one expresses several important ideas. Edit each sentence for conciseness, eliminating all unnecessary words (check Chapter 6 for suggestions on achieving conciseness). Remember, you are trying to pack as many important ideas as possible into a restricted number of words; so make every word count.

The short article reprinted here is followed by rough and polished drafts of an abstract and a summary of the article. Read the article, then study the editing process used in developing the polished abstract and summary.

Unlocking the Genetic Mystery*

"Do you want a boy or a girl?" is the question most frequently put to expectant parents. "It doesn't matter as long as it is healthy," is the usual response. It is also the usual outcome—well over 90 percent of babies born in this country are healthy and normal in all regards.

For those couples, however, who fall in the so-called "high-risk" category, whether because of age or family medical histories, the concern about a healthy baby is even more poignant. Now, thanks to a decade of tremendous progress in prenatal testing and, especially, clinical diagnostic techniques, these couples can be informed about the health of their unborn child. The vast majority of these diagnoses give the reassurance of a normal fetus (more than 95 percent of amniocentesis diagnoses do not contain the feared-for results), and the relief of having an anxiety-free pregnancy.

Prenatal Testing

The two major tools of prenatal testing are amniocentesis and chromosome banding. While amniocentesis has been with us more than 10 years, the recently developed technique of chromosome banding, developed initially in 1970, has made amniocentesis a much more sophisticated diagnostic procedure.

Amniocentesis, performed by a qualified obstetrician between the 15th [and] 18th weeks of pregnancy, involves inserting a hollow needle into the uterus

---

*"Unlocking the Genetic Mystery" by Judith M. Shapiro, M.P.H. Reprinted with permission of Mothers' Manual Magazine. © Vol. 16, No. 1, 1980.

through the abdomen to withdraw a small amount of the amniotic fluid surrounding the fetus. It is usually preceded by ultrasound examination, a painless procedure using high-frequency sound waves to transmit a picture of the fetus onto a television screen. With this picture the physician can precisely locate the fetus and the placenta to avoid touching them with the needle and to find the largest free area of amniotic fluid. (Ultrasound is also used to detect certain severe malformation in the developing fetus, multiple pregnancies, and fetal age.)

Once the amniotic fluid is obtained, it is sent to a laboratory where fetal cells are removed and allowed to multiply in a nutrient bath. After a few weeks the cells are analyzed for chromosome abnormalities or inherited disorders of body chemistry. Chromosomes are microscopic packages of vital hereditary information. Normal humans have 23 pairs of chromosomes in each of their body cells, including one pair of sex chromosomes—two X chromosomes in females, or an X plus a Y in males.

Chromosomes come in different shapes and sizes. When they were first looked at in the late 1950's and 1960's, it was hard to distinguish some of them from others. The chromosome banding techniques, a process through which chromosomes are stained with chemicals, was the vital breakthrough which enabled technicians to see differences and similarities much more clearly. This revealed chromosomal causes of many birth defects whose origins had previously been a mystery.

The most common serious chromosome abnormality is Down's syndrome, detectable only in some cases before chromosome bonding. Most children with Down's syndrome are born with an extra number-21 chromosome in every cell of their body and therefore have 47 chromosomes instead of the normal 46. These children exhibit varying degrees of mental retardation and usually have other health problems.

There are a wide range of inherited disorders of body chemistry that cannot be identified as chromosome defects can by bonding techniques, but reveal themselves as chemical abnormalities in amniotic fluid cells. Today, about 100 such defects are detectable prenatally. "But most of these disorders are rare and are tested for only if there is a specific reason for suspicion, such as family history or screening of carriers," explains Sara Finley, M.D. of the laboratory of Medical Genetics, University of Alabama in Birmingham.

Tay-Sachs disease, which disables and kills within the first five years of life, is the best-known example of a disorder which can be prevented by genetic screening of adults who may carry the abnormal gene. One in every 25 Jews of Eastern European descent carries it, and if two carriers marry, there is a one in four chance with each pregnancy that the child will be affected. Carriers for this disease can be identified by a simple blood test, but for a woman at risk, who is already pregnant, prenatal diagnosis is possible.

A third major birth defect for which amniocentesis is useful is an open malformation of the skull or spine, called a neural tube defect. Its effects include stillbirth, death in infancy, brain damage and paralysis. It can be detected by measuring the amount of a specific substance, called alpha-fetoprotein, in the amniotic fluid. Fortunately, the neural tube defect is rare; and although it is not hereditary, the test for it is performed primarily on women who have had a child or close relative with the defect.

Since it is virtually impossible to screen all pregnant women by amniocentesis, new testing procedures are being searched for. These would be to determine if there are indicators that the woman should indeed pursue amniocentesis. In the mid-1970's a few laboratories in this country and around the world began to explore the possibility of measuring the level of alpha-fetoprotein in the mother's blood as an initial screen.

"On the basis of the blood levels, we could focus down on a high-risk subgroup within that large, unselected pregnant population," reports Dr. James Macri, director of the Neural Tube Defect Laboratory at the State University of New York in Stony Brook.

Researchers are hopeful that in the 1980's this type of testing will be implemented via a network of regionalized centers. Ideally these would not only provide the service to a large population of pregnant women, but also have the necessary associated services, such as ultrasound, amniocentesis, highly reliable laboratory services and professional counselors.

Why Genetic Counseling

Many couples come to a genetic counselor because of a family history of birth defects which they believe to be hereditary. In these cases prenatal diagnosis should be preceded by counseling with a person trained in medical genetics. Often the fears of these couples are allayed and prenatal diagnosis is not recommended.

"All patients have different ethnic, moral and ethical backgrounds, as do most counselors. Counseling should be a vehicle of transmission of scientific knowledge at a level that can be understood by each patient. It should not be used to direct a decision which has moral or ethical implications," notes Garver. In the case of amniocentesis results that do show a defect, the affected couple is referred for further counseling to explore their options.

A Success Story

Today there may be positive options thanks to the many research scientists who are actively involved in devising in-utero treatment for some birth defects. To date, one case stands out in the medical journals. It concerns a child named April Murphy who was born with a rare vitamin metabolism defect called methymalonic adidemia (MMA). April's older sister had died of this shortly after birth and when her mother became pregnant again, she was offered prenatal diagnosis.

Test results showed that her second baby would also have MMA, but that the family's form of the disorder was responsive to vitamin B_{12}. It was decided to give Terry huge intramuscular injections of the vitamin—5,000 times the normal adult dose. The idea was to force the vitamin across the placenta to saturate both the mother's and the unborn baby's tissues, thus preventing possible prenatal damage.

Although April is kept on a controlled B_{12}-supplemented diet, in every other respect she is a healthy 6-year-old. And she has a new baby sister who was shown to be unaffected when Terry again had amniocentesis.

Looking Ahead

In spite of the rapid advances in prenatal testing, many genetic disorders remain undiagnosable prenatally. A link, however, may exist between the gene that causes such diseases and another gene whose product can be detected, making an indirect diagnosis possible. Gene mapping (determining where certain genes are located in a specific part of a chromosome) is a new scientific tool for researchers actively investigating classic hemophilia—an x-linked disorder—and juvenile diabetes, among others, for potentially useful gene linkages. Success will have a great impact on the future of clinical genetics.

It is hoped that major advances also will occur in this decade in the prenatal diagnosis of sickle cell anemia and cystic fibrosis. Sickle cell anemia, a blood disorder that affects primarily people of African descent, can presently be detected by a complicated and somewhat risky procedure of testing fetal blood. Researchers in California are now testing a new technique to diagnose the disease in amniotic fluid cells; this would make diagnosis safer and much more widely available.

Cystic fibrosis, the most common severe birth defect of body chemistry among Caucasians, is a hereditary disorder that appears in childhood and causes chronic lung infections, digestive problems, growth retardation and often death in young adulthood. Although scientists have tried to find a prenatal test for this disease, none has worked so far. But many experts believe that work going on at the present will eventually lead to a prenatal test for the disease as well as a carrier detection test.

While amniocentesis is a proven, safe, medical procedure, it is nevertheless "invasive"—that is, it involves entering the fetal environment. Consequently, its use is limited to women considered as potentially at high risk of giving birth to children with serious, prenatally detectable defects. To simplify prenatal testing—giving the same information as amniocentesis without invading the uterus—research is currently under way at Stanford University to develop a simple blood test that can also be used to determine fetal abnormalities. Dr. Leonard Herzenberg has found that a few fetal cells are present in the maternal bloodstream as early as the 15th week of pregnancy. These can be separated from maternal blood cells by using a very high-speed fluorescence-activated cell sorter. This technique, when perfected, could lead to widespread screening of pregnant women.

The future possibilities of medical genetics are exciting, but says Arthur J. Salisbury, M.D.-vice president for medical services of the March of Dimes, "We are faced with the disturbing fact that, despite the increased number of families who are helped by counseling services, 80 percent of the people who could be helped are not being reached."

One way to reach these people is to increase the number of "satellite clinics," staffed by genetic experts from major medical centers. These clinics, which are held periodically in community hospitals or public health facilities, bring modern genetic information to people who would have difficulty traveling long distances to a regional medical center.

The major problem in the future of clinical genetics is in funding. During the 1970's the March of Dimes had been the major source of funds to clinical centers in the United States which practice medical genetics. Currently the federal government is beginning to recognize the importance of genetic counseling services. The National Genetic Disease Act, which was passed in 1978, received an appropriation of $4 million; an additional $4 million has been tentatively added by the Congress to bring total federal funding to $8 million by 1979–80.

"Genetic services development in the 1980's is going to be strained financially unless there is more federal or other funding," adds Dr. Salisbury. "It's not so much that more major genetic centers are needed, but the increase in demand is so critical that they need more personnel and should be seeing more patients."

The gift of a precious life in health, prayed for by all parents, is the goal of hundreds of genetic scientists. Much has been done and, with the nation's support, more will be done during the next 10 years in solving one of life's greatest mysteries—life itself. ■

Rough Draft
Abstract

The article discusses what amniocentesis is and how it is performed. It then explains chromosome banding and how it and amniocentesis are used to detect genetic abnormalities. Some of these disorders, such as Down's syndrome, Tay-Sachs disease, and neural tube defect, are covered briefly. It then discusses who needs genetic counseling and gives a case study of one successful instance. Then some still undiagnosable conditions are discussed. Next, research into simpler tests was mentioned. Funding problems and potential finished up the article.

Final Draft
Abstract

Amniocentesis and chromosome banding are defined, then their use to detect genetic abnormalities is shown. Next, major disorders that can be diagnosed by these procedures are explained briefly, as are some disorders that are as yet undiagnosable. Future prospects for simpler tests accessible to more women are investigated. Finally, funding prospects and problems are presented.

Rough Draft
Summary

Ninety percent of all babies are healthy. Amniocentesis and chromosome banding can help many of the other 10 percent.

Amniocentesis is performed by a qualified obstetrician sometime between the fifteenth and eighteenth weeks of pregnancy. A hollow needle is inserted into the uterus to withdraw a small amount of the amniotic fluid surrounding the fetus. Ultrasound is used to locate the best place to insert the needle.

Fetal cells are allowed to multiply in a nutrient bath for a few weeks, then analyzed for chromosome abnormalities or inherited disorders of body chemistry.

Chromosome banding, a process in which the chromosomes are stained, was a vital breakthrough that made possible detection of chromosome abnormalities such as Down's syndrome. Most children with this disorder have an extra chromosome in each cell and various health problems including mental retardation.

A wide range of body chemistry disorders can be revealed through detection of chemical abnormalities. About a hundred such defects can currently be identified this way. Tay-Sachs disease is the best known example. It disables and kills before age five. It is identified through an abnormal gene carried by one in every twenty-five Jews of Eastern European descent. A child born to two carrier parents stands a one in four chance of having the disease.

Amniocentesis is also effective in diagnosing an open deformation of the skull or spine called neural tube defect. Its effects include stillbirth, death in infancy, brain damage, and paralysis. It can be determined by measuring a substance called alpha-fetoprotein in the amniotic fluid. The disease is rare so tests are performed only on women with a family history of the disease.

Efforts are now underway to find prenatal tests to identify cystic fibrosis, the most common severe birth defect among Caucasians.

Although amniocentesis is a proven, safe procedure, the fact that it is invasive of the uterus limits its use to high-risk women. Many doctors are trying to devise simpler prenatal tests that could be used for widespread screening.

The major problem in the future of clinical genetics is lack of funding. The March of Dimes has heretofore been the major source of funds, and the federal government is considering some funding, but more is needed. Over 80 percent of the women who could use help are not getting it.

Final Draft
Summary

Amniocentesis and chromosome banding can help the parents of the 10 percent of American babies not born perfectly healthy. Amniocentesis, performed by an obstetrician between the fifteenth and eighteenth weeks of pregnancy, involves withdrawing a small amount of amniotic fluid.

After multiplying for a few weeks in a nutrient bath, fetal cells are analyzed for chromosome abnormalities or inherited body chemistry disorders. Chromosome banding, a vital breakthrough involving chromosome staining, makes possible identification of many abnormalities such as Down's syndrome. Most children afflicted with this disorder have an extra chromosome in every cell and suffer numerous health problems, including mental retardation. About a hundred other defects can be found through detection of chemical abnormalities. Tay-Sachs disease, the best-known example, disables and kills children before age five. One in twenty-five Jews of Eastern European descent carries the abnormal gene, with two carrier parents standing a one in four chance of having a diseased child.

Amniocentesis can also help to diagnose neural tube defect, an open deformation of the skull or spine that causes paralysis, brain damage, death in infancy, or stillbirth. It can be detected by measuring alpha-fetoprotein in the amniotic fluid. Tests for this rare disease are given only to women with suspect family histories.

Researchers are now trying to find a test for cystic fibrosis, the most common severe birth defect among Caucasians.

Although amniocentesis is proven safe, it is invasive of the uterus, so it is used only on high-risk women. Researchers are trying to devise simpler tests to permit widespread screening.

Because of a fund lack, only 20 percent of the women who need help get it. The March of Dimes is the main source of funds, with the government contributing some money and considering appropriating more.

Main Text

As its name implies, the main text is the largest portion of the report, often comprising up to 90 percent of the entire document. It contains the same three sections that any well-prepared composition does: in introduction, a body, and a terminal section.

1. Introduction ● The introduction prepares the reader for the main discussion, providing a context for interpreting and judging the new information you will present. Begin by making clear your purpose in doing the study or project on which the report is based. Answer such questions as: What did I hope to accomplish? Why did I want to accomplish it? To what use do I hope to put my results? Is the report part of a larger study or a long-range project? Am I working on it independently or as part of a team? Is it being done strictly for my own or for some other organization? Is it a follow-up to earlier efforts? No introduction will answer all of these questions, and some of the questions will not apply to every situation. A general rule is to give a full explanation of the nature of your project or study.

If the project or study is based on a particular theory or hypothesis, state it clearly. Also give any formulas, equations, or technical data the reader should know before reading the main discussion.

If no preface or acknowledgment section is included, your introduction can acknowledge someone's assistance or comment about the report's structure or style. This is also an excellent place to indicate one or two printed sources of information. If a number of sources are used, you should include them in a bibliography or list of references at the end of the report.

Double-check the introduction to make sure you have not omitted anything necessary. Edit it for clarity and conciseness. Do not worry if your introduction is quite short. It is perfectly acceptable for a fifteen-page report, for example, to have an introduction that runs less than one full page.

2. Discussion, or Body ● The longest single portion of the typical report is the discussion (sometimes called the *body*), in which most of the new information is presented. If the report is based on a study, the discussion contains the information gathered and examines, point by point, the topic under study. If the report is based on a project, the discussion traces, step by step, exactly what was done. A later section discusses this part of the report in more depth.

3. Terminal Section ● Because this section will probably be read by more people than any other part of the main text, it must be especially effective. A good terminal section consists of three logically sequential parts: findings or results, conclusions, and recommendations.

Two qualities are essential to an effective terminal section. It must be thorough, and it must be logical. Every recommendation made must be clearly based on one or more conclusions, which in turn must be logically based on findings or results. Readers who wish to verify a recommendation before acting on it should be able to trace it back through the conclusions and into the findings.

Present your findings or results in raw form. Use tables, lists, and similar illustrative devices whenever possible. Then use conclusions to evaluate and show the significance of the raw data. Include all your findings, even those you consider to be unreliable, insignificant, or invalid. But be sure to explain why you are rejecting them.

If your report is based on a study in which your purpose was to gather information, your discussion section will, in effect, constitute your findings; so the terminal section can begin with your conclusions. Yet the principle here is the same—you state information, explain and interpret it, then offer recommendations.

Back Matter

Unlike short reports and most other written technical communication, long, formal reports seldom end with the recommendations. Most contain at least some back matter, and some may contain more back matter than main text. Four types of entries may be included as back matter: lists of references or bibliographies, appendixes, glossaries, and indexes.

1. Reference List or Bibliography ● If you use direct quotations or paraphrased ideas from other sources, you are ethically (and often legally) obligated to give appropriate credit. Certainly you must enclose in quotation marks any material you use word-for-word from another source. And you must also identify the author and the specific document from which you are quoting. Even if you do not quote verbatim, you must identify the author and document from which you borrow an idea, a fact, or a figure.

When your report contains only one or two references to or quotations from another source, you can given appropriate credit by mentioning the author's name and the document title in the text. When you rely heavily on one or two reference sources as guides for the entire project, you can give credit in the acknowledgment section or in your introduction. Only when you make numerous references to several (at least three or four) different sources do you need formal documentation. Such documentation can take either of two forms: footnotes and bibliography or reference numbers and a reference list. Many occupational areas prefer reference lists because they are easier to compose than bibliographies and footnotes. However, check to see which technique is preferred by a majority of people in your field. Chapter 18 gives suggestions and models for using both systems. When you use either, make it your first item of back matter.

2. Appendixes ● If some of the material you want to use in the report does not fit directly into the main text, include it as an appendix. Appendixes may contain any type of material relevant to the main text. Illustrations, samples, related information, computer printouts, data collection sheets—if they will physically fit into the report—can be included.

A key word to help you decide what to include in the appendix is *supplementary*. Appendixes supplement the main text, adding extra information for clarification or support. They should not contain anything absolutely essential to a thorough understanding of the text. If your reader will be required to read an appendix in order to comprehend something in the main text, move the material from the appendix into the main text. Never force your reader to refer to an appendix while reading the main text.

For each appendix, provide a title and an appendix letter. For instance, you might have "Appendix A: Customer Questionnaires," "Appendix B: Completed Program Flow Sheet," and "Appendix C: Sample Program Printout." There is no established limit for the number of appendixes you can include. One report may have no appendixes; another may have more material in the appendixes than in the main text.

3. Glossary ● In very long reports written for nontechnical readers, it is sometimes necessary to include a glossary, an alphabetized compilation of definitions. (See Chapter 3 to brush up on definitions.)

Do not use a glossary if you have only a few terms to define. You can include four or five working definitions in the text, using parentheses or explanatory notes. You can also include a one- or two-page list of definitions as a preliminary part of the report body. This technique allows readers to learn the definitions of key terms before encountering them in the text. It also saves readers the trouble of looking for a glossary in the back material, where it can easily be overlooked.

Glossaries are very helpful in reports that are to be read both by highly technical subject-area experts and by semitechnical or nontechnical managers and administrators. Those who need the definitions can use the glossary, yet it need not interfere with those who already are familiar with your terminology.

4. Index ● The least-used portion of the back matter is the index, an alphabetical list of important topics covered and the page numbers on which they are discussed. Use one only if the report is quite lengthy (at least fifty and usually one hundred or more pages), and then only if you are certain that the report will be used for reference. Should someone want to look up specific items over a period of time, the index will be useful; otherwise, it will probably be ignored.

Constructing an index is more time consuming than difficult. Go through the main text, jotting down each important item and its page number on separate three-by-five index cards. Because you will list all items that the reader might want to locate, include every item you are in doubt about. When your list is complete, alphabetize the stack of cards. As always, double-check your work to make certain that you have not forgotten something. Now type up the finished index from the alphabetized cards.

The Discussion, or Body

As the preceding section noted, the discussion, or body, of a long, formal report may comprise as much as 80 to 90 percent of the material. Fortunately, it is relatively easy to construct.

Each report, like most other pieces of writing, is unique; yet most fall into two basic categories: the study report and the project report. The body of each is constructed according to a basic pattern: the project report body is usually a technical process or narrative; the study report body is usually a technical classification or analysis. Each can be developed with the basic techniques explained in Chapter 4.

The Study Report ● A study report is essentially a presentation of information. The writer obtains certain data or studies a certain problem or situation, then reports the information. When you do such a study, put the information you obtain into the body of your report, then interpret and draw inferences from it in the terminal section.

The key to a well-organized body is the outline. Develop a rough outline, classifying your subject into a few general subtopics. Then gather information according to the outline, making certain to develop each subtopic fully. Your rough outline is not sacred. Do not hesitate to add, delete, or combine subtopics. Use the outline only as a means of organizing your information in order to facilitate writing it up later.

Once you have your material gathered, divide each section carefully, using the organizational patterns discussed in Chapters 3 and 4. One section might be descriptive, another comparison-contrast, and another process analysis. If no obvious pattern appears, divide the section according to some logical principle of classification. The important thing is to develop each section according to the pattern that makes sense for it.

Now you can develop a rough draft for each section, using the techniques appropriate for each pattern. Deal with one section at a time, writing it as though it were an individual composition. If some sections are too long or complex, you can subdivide them further and handle one subsection at a time.

Complete rough drafts of all the sections before revising. Check to make sure that all relevant material is covered and placed in the appropriate section, and that sections are placed in the proper order. Then revise the draft, making certain that the sections fit together coherently.

The Project Report ● Many long technical reports are accounts of jobs or tasks performed. As common sense indicates, the most effective pattern for such a report is a chronological narrative. Your introduction can explain the purpose of the project and how you came to be working on it. The body can begin with the planning stage of the project and proceed step by step through the completion of work. The terminal section can show the results of the project and offer conclusions and recommendations.

If you know before beginning a project that you will have to write a report about it, take careful notes every step of the way. That will save you a great deal of time and effort later on when you are reconstructing every significant phase of the operation for the report.

Collect and examine your notes before you begin to work on the body of your report. Beginning with the initial stages of the project, develop the outline, using one section for each major phase of the operation. Then subdivide each section of the outline as shown in Chapter 5.

Do not try to narrate the project as a whole; concentrate on describing each phase clearly. Be thorough, mentioning exactly what you did and when you did it. If methodology is important, explain how you did things. If you had unexpected problems or success, explain your procedures carefully.

Complete a rough draft of each section, polish it, and check for coherence.

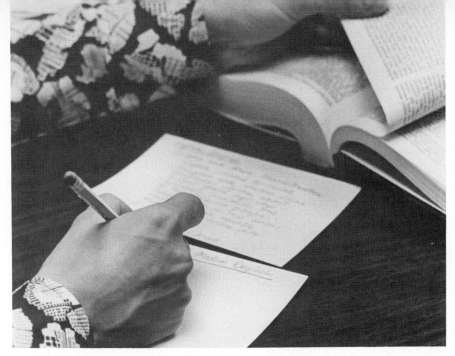

If you take careful notes while researching a project report, you will spend less time developing the outline and writing the final paper.

Special Qualities of Formal Documents

The long, formal report, as well as other formal documents, must be polished especially carefully; it should be as flawless as possible. Here is a checklist of special considerations to keep in mind when preparing a long, formal document.

1. Use the best possible format. If you are given a format to follow, use it faithfully. If you devise your own, be sure that the parts you include are appropriate and that they are in the proper order.
2. Differentiate fact from judgment. Your facts must be accurate, your judgments sound and well supported.
3. Be thorough. Include all relevant data. Write as concisely and straightforwardly as possible. Do not pad.
4. Use illustrations liberally. Tables are especially useful. Use them and any other type of illustration that will help you communicate more clearly, concisely, and forcefully. Do not, though, use illustrations merely for appearance or to attract attention (see Chapter 7).
5. Proofread your finished product several times. Use the handbook portion of this text and your dictionary to check for mechanical problems and misspellings.
6. Prepare your manuscript carefully. It should be well typed, with no strikeovers or obvious corrections. Use high-quality paper and a typewriter with clean keys and ribbon. If the manuscript is going to be printed or otherwise duplicated, see if there are any special problems that need to be dealt with before you submit it.

Before writing a formal report, examine the model report in Figure 10.3.

Valencia Community College

West Campus

Orlando, Florida

Report on

DESIGN AND PLANTING PROCEDURE

FOR DISPLAY OF ANNUALS

IN MECHELBURGH TOWNSHIP PARK

Date of Submission: June 4, 1980
Submitted to: David Fear, Instructor
Submitted by: Richard Jones
Required for: English 157

Figure 10.3 Model Formal Report

TABLE OF CONTENTS

Page

1.0 SUMMARY

As part of the overall beautification project, beds of dwarf annuals were set out in the Mechelburgh Township Park. Dwarf hybrid strains provide quick, bright color with minimal maintenance. The procedures undertaken for preparing and planting beds may serve as a guide for future planting.

2.0   INTRODUCTION

This project was part of an overall beautification program
undertaken by Abigail Landscaping for the Township of Mechelburgh.
The purpose of the program was to restore the one-third acre park
on the ground between the county and township office buildings
on Hampton Street.

One stipulation of the Mechelburgh Township Council was that
the plantings in the park include areas of bright color,
preferably on a patriotic theme. It was therefore agreed
between our representative, Timothy Reiner, and representatives
of the council that the best way to achieve quick color would
be to include plantings of annuals as part of the restoration
project.

Because the annuals will last only one season and the
council will have to determine from year to year whether to
continue the display or to turn to more permanent plantings,
this report has been prepared apart from the overall report on
work within the restored park. Because the project required
investigation of the availability and costs of annuals in our
area, it may serve as a current guide for other projects.

2

3.0 BODY

3.1 Definitions of Terms

1. Annual: a plant that grows from seed, flowers, produces
   seeds, and dies in one season.

2. Contact-kill herbicide: a chemical compound that enters
   and kills plants through the leaves without contaminating
   the soil.

3. F-1 hybrid: a first-generation hybrid whose parents are
   different pure-bred strains; such plants are generally
   more vigorous than other hybrids.

4. Humus: decayed organic matter, a vital component of good
   topsoil.

5. Root ball: the thickly matted roots and soil filling the
   container of a healthy plant.

6. Slat: a plastic tray designed like an egg carton with
   seventy-two separate compartments which serve as pots
   for the seedlings.

7. Vermiculite: an artificial growing medium of expanded
   mica used to retain moisture in soil mixtures.

3

3.2 Location and Design of Beds

Since the purpose of this project was to add color and
attract attention to the newly restored park, first consideration
was given to locations bordering on Hampton Street. (Figure 1
shows the layout of the park.) The eight-foot-wide strip between
the street and the walkway, having almost full sun and high
visibility, offered an ideal planting area (Beds A and B).
Although establishing beds here would require removing well-
established grass, the sod taken up could be used in bare areas
near the County Building.

The area surrounding the Statue for the Missing in Action,
located near the Township Building, was also selected for annuals
(Bed C). Here there were existing beds which, though overgrown,
could easily be restored. Colorful annuals could draw attention
to the memorial despite its rather obscure location in the
park.

Strong consideration was given to the area surrounding the
fountain at the northwest corner of the walk, but this location
was rejected because of the strong possibility that visitors
looking at the restored fountain would step into the beds and
destroy the plants. More durable plantings would be used in this
area. However, the circular theme of the fountain was picked up
for a bed in the southeast corner of the park (Bed D), providing
balance for the fountain and accentuating one of the entrances

4

Figure 1   Ground Plan of Park Showing Location of Beds for Annuals

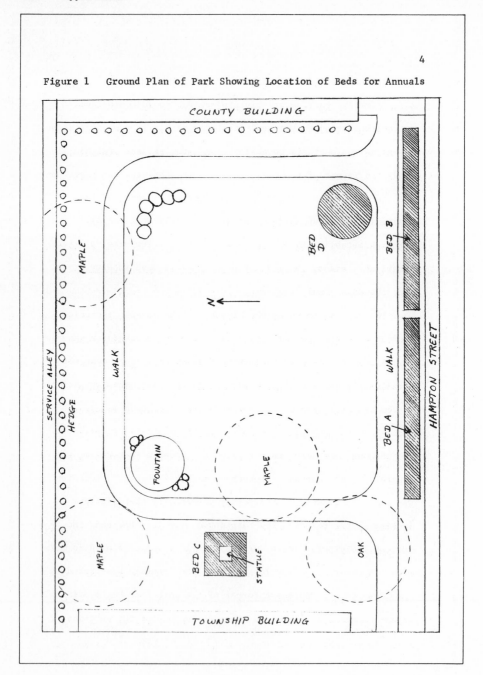

5

to the park. A smaller, square bed already established at this
point would be enlarged to a circle sixteen feet in diameter.

To ensure effective color separation, the beds were designed
for planting in blocks or lines of no less than two rows of each
color. Figure 2 shows the detailed arrangements of the beds.

3.3 Selection of Annuals

After a preliminary study of the different annuals available
in quantity locally (listed in the Appendix), I selected the
Dwarf Cascade Petunia for Beds A, B, and D. It is one of the
newer F-1 Hybrids, exhibiting uniform growth, wide color selection,
and superior flower production. The plant grows to 10 inches in
height with many strong basal branches and flowers averaging $3\frac{1}{2}$ to
5 inches. Using the dwarf strain will substantially reduce
maintenance, as these plants tend to grow low and spread out.
Low growth eliminates the need for staking, while spreading
helps to control weeds by forming a solid mat of foliage in the
bed.

Bed C, shaded by the Township Building and by the large trees
at the west end of the park, would not receive sufficient sunlight
for petunias. The only readily available annuals that could be
counted on to flower over a long season in these conditions were
begonias, impatiens, and vinca. None of these produces a blue
flower, and vinca, otherwise highly satisfactory, lacks a true
red. I selected the impatiens as the most likely to bloom well

6

Figure 2   Detailed Design of Beds

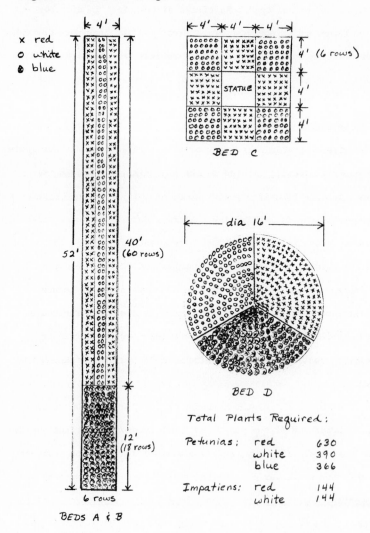

x red
o white
● blue

BED C

dia 16'

BED D

Total Plants Required:

Petunias: red   630
             white   390
             blue   366

Impatiens: red   144
             white   144

52'   40'
       (60 rows)

12'
(18 rows)

6 rows

BEDS A & B

and require little maintenance. After discussion with Mr. Reiner,
I settled on a display using only red and white. I selected the
Imp Series hybrid, which grows to 15 inches and produces especially
large flowers for the species from midsummer to frost.

3.4 Preparation of Beds

I began the preparation of Beds C and D using a contact-kill
herbicide on all vegetation in the existing beds. A single application of Paraquat CL, a nonselective herbicide, was adequate to
clear the beds of all growth. A solution of 1 quart to 40 gallons
was sprayed on the beds. After a few days, the beds were raked
off and trimmed to the correct size and shape.

Beds A and B were carefully staked off, and the sod crew
removed the sod. The beds were spaded to a depth of 12 inches
and all large stones, pieces of concrete, and other debris were
removed. Next, humus was added to all beds in the form of a
compost consisting of peat, vermiculite, sphagnum moss, manure,
and bone meal. After the mixture was spread, it was thoroughly
worked into the top 8 inches of dirt watered heavily for the
next 2 days.

The final step in preparing the beds was to apply $1\frac{1}{2}$ pounds
per 150 square feet of Dowfume MC-2, a fumigant, to ensure a
total kill of all nematodes, weed seeds, and fungi in the soil.
Since the fumigant's active ingredient is methyl bromide, a gas,

8

it was necessary to tent the beds with plastic sheeting before
application. Although this required additional labor, it resulted
in sterile beds that increased the survival rate of seedlings and
eliminated future weed growth. After a 24-hour exposure period,
the plastic was removed and the beds were watered moderately for
the next week.

The entire preparation required 42 man hours, of which at
least 20 would not be necessary for subsequent plantings in these
beds.

3.5   Planting Procedure

The planting procedure began with a final tilling and leveling
of the beds immediately before the rows were laid out. Row
stakes were then positioned to guide in punching the planting
holes. A planting stick was used to make uniform holes 1½ inches
square and 2½ inches deep. As each color section or row was
punched, a color stake was pushed in identifying the correct color
to be planted in the row. As soon as a major section of the bed
was ready to plant, the slats were unloaded and plants of appropriate
color were placed at the head of each row. When the plants were
removed from the slats, the root ball of each was dipped into a
very dilute solution (½ ounce to 10 gallons) of Truban fungicide
to prevent root rot. The plant was then placed in the hole and
dirt pressed gently around the roots. After the bed was completed,

9

it was watered heavily. This process was repeated until all the
beds were planted. The planting procedure required 12 man hours.

3.6 Care After Planting

Although the beds required watering every two days for the
first ten days after planting, the hours needed for future
maintenance will be kept to a minimum. Watering will be taken
over by township personnel. Bimonthly cultivation between the
rows and feeding to ensure continued growth will be undertaken
by Abigail Landscaping, as contracted; and I will be responsible
for periodic visual inspections to watch for developing pest
problems and to check the general health of the plants.

10

4.0   TERMINAL SECTION

The plants appear to be doing very well after two weeks'
care and should certainly provide the display of color requested
by the Township Council. Although the preparation and planting
required a number of steps separated by days of watering, the
project was easily finished on schedule.

This report may serve as a rough guide for future plantings
of annuals in the park. Much of the preparation work will not
need to be repeated, although treatment of the beds with fumigant
every other year will help keep down maintenance.

11

APPENDIX

Annuals Available in Quantity, Spring 1980.

Species	Variety	Average Height (inches)	Color[a]							Season[b]			Cost per Slat
			W	R	B	P	L	Y	O	Sp	Su	F	
Aster	Pepite	12	x	x				x	x		x	x	$9.50[c]
Begonia	Semperflorens Dwarf	8	x	x		x	x	x	x	x	x	x	16.00
Dahlia	Dwarf Bedding	16	x	x		x	x	x	x	x	x	x	18.25
Impatiens	Imp Hybrid	12	x	x			x	x	x	x	x	x	12.00
Lobelia	Cascade	5	x	x	x						x	x	8.00
Petunia	Double Hybrid Multiflora	14	x	x	x	x	x	x	x		x	x	9.50
	Dwarf Cascade	10	x	x	x	x	x		x	x	x	x	8.75
	Grandiflora Double Hybrid	14	x	x	x	x	x			x	x	x	9.00
	Multiflora Joy	12	x	x	x	x	x			x	x	x	7.95
Portulaca	Sunkiss	8	x	x				x	x	x	x	x	8.25
Verbena	Regalia	10	x	x			x	x			x	x	7.50
	Amethyst	10				x					x	x	11.95
Vinca Rosea	Border Type	12	x			x	x	x	x	x	x	x	7.25
Zinnia	Paint Brush	16	x	x			x	x	x	x	x	x	7.95
	Thumbelina	8	x	x			x	x	x	x	x	x	7.50

[a]White, red, blue, pink, lavender, yellow, orange.

[b]Sp (spring) blooms May–June; Su (summer) blooms July–mid-August; A (autumn) blooms mid-August–September.

[c]Currently available only in slats of mixed colors.

Semiformal Reports

As mentioned before, technical reports come in all sizes, shapes, and descriptions. The preceding chapter offered models of and suggestions for one extreme—the memorandum. The first section of this chapter covered the other extreme—the long, formal report. Between the two lies a vast array of reports, for convenience' sake referred to as *semiformal* reports.

These reports run from a minimum of three or four pages up to twenty or thirty pages, with five to ten most common. They include some, but not all, of the trimmings associated with formal reports. In scope they are more like memorandums, concentrating on giving necessary information to a specific reader.

Use a semiformal approach when you want something more impressive than a memorandum. Use it also for midlength reports aimed at a specific reader, but check to make sure that company practice does not dictate a more formal approach. Follow standard practice within your organization and use these guidelines to help you to decide which items to include:

1. Use a title page unless it is not done in your organization. (See pages 184–185 for suggestions.) Approval and distribution pages are rarely needed.

2. Do not use a preface. Save it for the most formal documents.

3. Use a brief table of contents for a midlength report; do not bother for shorter reports. Five or six pages is probably a good cutoff point.

4. Use a list of illustrations only if you have three or more illustrations. You can put it on the same page as your table of contents.

5. Use an abstract or summary only with a midlength report. If you can choose between the two, keep in mind that a summary is generally more useful. Notice that the first model below includes a brief abstract on the title page. This is a fairly common approach. Notice also that it also takes the place of a table of contents.

6. Do not use a letter of transmittal. If you wish, you may include a brief memorandum of transmittal, especially if your reader is not expecting the report or may not be inclined to read it.

The main text consists of the same three basic sections as memorandum reports and formal reports: introduction, discussion, and terminal section. Figure 10.4 shows a model semiformal report, and Figure 10.5 shows a model memorandum of transmittal.

ARDAMAN ENGINEERING ASSOCIATES

West Highway 50

Orlando, Florida 32811

Report on: Sinkholes in Central Florida

Abstract: The properties and characteristics of limestone are
explained, followed by a discussion of fractures of limestone
as predictors. The three types of sinkholes--collapse, dolines,
and erosion--are defined and illustrated.

Illustrations: Figure 1 Primary Porosity

Figure 2 Secondary Porosity

Figure 3 Stages of Development of Collapse Sinkholes

Figure 4 Stages of Development of Dolines

Figure 5 Stages of Development of Erosion Sinkholes

Introduction

This report describes the development of three types of sinkholes common to Central Florida. Engineers have been studying the
phenomenon of sinkholes for many years, trying to understand why
they occur. Although it is a complex subject, a person does not
need a degree in geology to understand the fundamentals of sinkhole
development.

This report should help explain sinkholes and why they occur.
I gathered most of my information from technical reports issued by
Ardaman & Associates, Inc., of Orlando and from articles in the
Journal of the Geotechnical Engineering Division, American Society
of Civil Engineers.

Figure 10.4 Model Semiformal Report

Properties and Characteristics

Since the development of a sinkhole begins in the subsurface limestone formation, a good understanding of the properties and characteristics of this limestone is essential.

Limestone is made up of calcium carbonates. The biological ingredients include coral, shell fragments, and sand-sized remains of animal and plant life. When freshly deposited, these formations exhibit high void ratios from the pores between the grains and fragments. These pores are termed <u>primary porosity</u>. (See Figure 1.) Once the limestone hardens even small strains can cause cracks in the structure; sometimes this occurs even under the rock's own weight. These cracks are termed <u>secondary porosity</u>. (See Figure 2.)

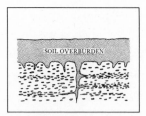

Figure 1 Primary Porosity

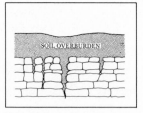

Figure 2 Secondary Porosity

Carbonates are soluble in water that is slightly acidic. As rain falls through the air it mixes with carbon dioxide, forming weak carbonic acid. Since this acid is a natural solvent for limestone, solution occurs as it passes down from the ground surface through the limestone. A number of factors affect the rate of solution. The more acidic the water, the greater the rate. The more circulation of water through the pores and cracks, the faster the rate of solution. The increased solution enlarges the pores, which in turn increases circulation, creating a self-aggravating process. As the pores get

larger the limestone structure is weakened. Therefore, limestone
is not an inert material; it is dynamic, changing with the environment.[1] This dynamic character leads to formation of sinkholes.

Fracture Patterns as Predictors

The development of sinkholes is expected to be concentrated
near the intersection of joints and fractures in the limestone.[2]
Geologists believe that a study of these features should provide
information on the possible locations of future sinkholes.

Since most of the rock formations in Florida are covered by some
type of overburden soil, the fractures in the rock are not visible.
However, their location can be deduced from examination of surface
features such as stream patterns and sinkhole alignments.[2] These
can be used to map lines along which sinkholes might be expected to
occur.

Types of Sinkholes

Collapse Sinkholes

Though not very common, collapse sinkholes happen suddenly and
are disastrous. The limestone formation is generally at or near the
surface. The sinkhole develops when a cavity in the limestone enlarges to the point where it collects water from smaller cavities,
increasing the solution activity in the large cavity. (See Figure 3.)
Many factors can initiate the failure, but the increased solution
activity is ultimately responsible.[1] An increase in rainfall or the
pumping of water from the ground for drainage can increase the weight
of the soil overburden, causing the cavity to collapse. The actual
failure occurs when the cavity can no longer support the weight of the

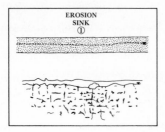 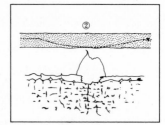

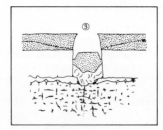

Figure 3 Stages of Development of Collapse Sinkholes

overlying soil and rock. The collapse of the cavity is followed by a drop of the soil overburden. The hole is generally steep-sided and rocky.

In January, 1969, four spans of a reinforced concrete bridge near Trapon Springs, Florida, failed because of a collapse sink. The collapse was so sudden that traffic could not stop, resulting in one death and several injuries.

Dolines

Two types of dolines are common in Central Florida: overburden dolines and rock surface dolines. They occur when the cavities and channels in the limestone formation extend upward to the top of the rock formation. The soils overlying the limestone are generally cohesionless, sandy soils. In the overburden doline, small portions of these soils are gathered up by the flow of water and carried down into the channels of the limestone like sand through an hourglass.[1]

In the rock surface doline, fragments from the surface of the limestone are gathered up by the flow of water and carried downward the
same way. In both types, the ground surface generally slumps over a
long period of time instead of falling suddenly. It takes approximately 5,000 years for the ground surface to drop one foot. This
process will continue as long as there is sufficient circulation of
water through the cavities in the limestone. (See Figure 4.)

Most of Florida's lakes are dolines that are hundreds of
thousands of years old.[3] Most of the shallow wet-weather ponds are
also examples of dolines.

Erosion Sinkholes

Erosion sinkholes, the most common type of sinkhole in Central
Florida, are very dangerous. Their development is similar to that

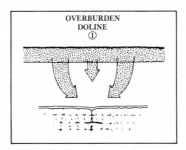
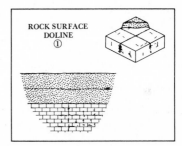
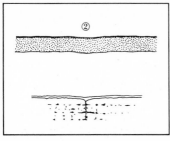
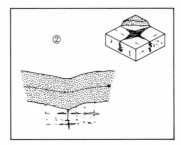

Figure 4 Stages of Development of Dolines

of dolines except that the overlying soils are cohesive and claylike.
As the water flows down from the ground surface, the soil gradually
flakes off above the rock surface, forming a cavity in the overlying
soils. This soil is carried down into the channels of the rock.
When the cavity becomes so large that the roof formed by the remaining soil can no longer support itself, a sudden collapse occurs. The
cavity in the overburden soil is generally domelike and wider than it
is high.[1] The roof remains intact when the failure occurs. If the
collapse is shallow enough, grass and trees can be seen growing in
the bottom of the hole. However, many times the soil drops completely
out of sight. (See Figure 5.)

Large amounts of rainfall after a long dry spell can cause this
type of failure. During a dry period the clay soils may crack as
they lose moisture; thus an increase in rainfall will increase the
seepage through the soil.

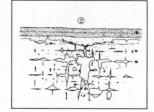
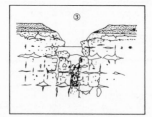

Figure 5 Stages of Development of Erosion Sinkholes

Erosion sinkholes are typically less than forty-five feet in diameter, but some are as big as two hundred feet.[4] In April, 1975, an erosion sinkhole approximately sixty feet in diameter engulfed the cafeteria at the Audubon Park School in Orlando, Florida. One that developed at a residence in South Orlando in May 1977 extended sixteen feet under the house.

Conclusion

After many years of research, the hows and whys of sinkhole development are pretty much understood. When and where they will occur is still a mystery. However, by conducting deep soil borings it can be determined whether conditions are favorable for sinkhole development.

Several methods are used to stabilize a potential erosion or doline sinkhole once its location is determined, but there are no guarantees. One method is to inject a cement mixture into the ground at the rock surface, sealing the cavity opening and therefore eliminating circulation. Sand must then be injected into any cavities that exist in the overlying soils. Once the site is stabilized, it is relatively safe to build a single-story structure on it. Another method is to drive concrete piles down to the rock surface and pour a concrete pad over the top of the piles. Then a structure can be built on top of this pad.

Whether a person is considering the construction of a single-story residence or a multi-story office building, I highly recommend a sinkhole investigation before a site is purchased.

References

1. Sowers, George F. "Failures in Limestones in Humid Subtropics."
   Journal of the Geotechnical Engineering Division, ASCE, Vol. 101,
   No. GT8, Proc. Paper 11521, August 1975, pp. 771-787.

2. Garlanger, John E. "Sinkhole Phenomena." Lecture at University
   of Florida, 1977. Dr. Garlanger is a geotechnical engineer for
   Ardaman & Associates, Inc., Orlando, Florida.

3. Ardaman, M. E., "Springs and Sinkholes: Blessings and Curses."
   Journal of Florida Engineering Society, November 1975, pp. 8-10.

4. Ardaman & Associates, Inc., Orlando, Florida. From technical
   reports, File Numbers 73-500A, 75-255, and 75-287.

MEMORANDUM

Ardaman Engineering Associates

To: Kathleen Coleman   Date: March 21, 1981

From: Robert Felter   Subject: Background report on
                                           Florida sinkholes

Attached is the background report on Florida sinkholes for you and

the other members of the drafting department. I hope it will give

you the working knowledge of sinkholes you need without miring you

down in technicalities.

We appreciate your interest in improving your department. Please

let me know if I can be of further help.

Figure 10.5 Model Memorandum Transmittal

Progress Reports

The progress report does simply what its name implies: it reports progress on a particular piece of work. It can be long and formal or short and informal. Regardless of length, formality, or type of work being reported, certain characteristics are common to nearly all effective progress reports.

Basic Structure

Like any effective report, a well-composed progress report has an introduction, a main body, and a terminal section. Since the progress report covers only part of a job, however, its introduction and conclusion are a bit different from those of the other reports we have been discussing.

1. Introduction ● The nature of the introduction to a progress report depends upon whether the report is the initial one in a series or a subsequent one. The introduction to the first progress report on a project is much like that of any other report. You explain why the project is being done and what it is intended to accomplish. You also mention who authorized or ordered the project and who is working on it. In short, you explain the background of the project.

Subsequent reports on a project need not give a complete background introduction. Because your reader has probably seen your initial report, you should use what is sometimes called a *transitional introduction*. It states in one or two sentences the purpose of the project, then gives a brief summary of work accomplished during prior reporting periods. For instance, if your report covers work done during May and your previous reports have dealt with work done during March and April, you should include a brief summary of work accomplished during these two months. This summary will provide the proper context for the description of progress made during May. Here is a sample transitional introduction:

```
                        1.0  Introduction

    Purpose:  This project is designed to determine the tasks performed
    by electronics technicians throughout our nine-state area, then to
    help area colleges develop a curriculum that teaches the skills
    needed to perform those tasks.

    Summary of earlier work accomplished:  During March we designed a
    task inventory, field tested it, and made needed refinements.  Dur-
    ing April we administered the inventory to two hundred working
    technicians.
```

2. Discussion ● The primary function of the discussion, or body, is to tell the reader what progress has been made during a given reporting period. To do this, you may construct either a chronological narrative or a topical discussion. Experienced progress-report writers often prefer the topical discussion, but either approach will work. In reporting your accomplishments, describe the tasks actually performed rather than the procedures you used.

If you have run into difficulties on the project, explain those problems clearly and fully. If the job is behind schedule or is not going the way it should, your reader will want to know the details. Be candid and objective. Avoiding the mention of embarrassing problems may have temporary advantages, but doing so is too big a gamble to take. Eventually the reader finds out and is much more angry than if he had been kept well informed throughout.

Here is an effective discussion section from a progress report:

```
                    2.0  Discussion

   Tasks accomplished:  1.  Collated completed inventories.
                        2.  Completed item analysis of inventories.
                        3.  Developed fifty objectives.
                        4.  Began developing criterion references.

   Methods used:  Professor John H. Wilkinson of Westmoreland Techno-
   logical Institute did our statistical analysis.  We did the rest
   ourselves.

   Difficulties encountered:  Wilkinson was ten days late with the
   completed item analyses.  Then several team members wanted to include
   objectives not justified by the statistics.  The ensuing discussion
   cost us two more days.
```

3. Terminal Section ● Because the project is incomplete, most progress reports cannot offer the conclusions and recommendations found in other project reports. Only the last progress report (often called a terminal report) can include firm conclusions or recommendations. Other progress reports, though, should include what are sometimes called *prophetic conclusions.* A prophetic conclusion can include any tentative conclusions you have reached about the entire project, but its main concern is with the remainder of the project.

Begin by telling the reader exactly what part of the project you are going into during the next reporting period. Let him or her know what to expect in the next report. Then project past the next reporting period to the job's completion. Are you on schedule? How far ahead or behind? Do you expect to get back on schedule? What do you plan to do to catch up? Do not mislead the reader about staying on schedule. Do not promise that you will be back on schedule next time if you know that it is unlikely. In fact, the final part of your prophetic conclusions should be direct recommendations for changes in schedules, methods, personnel, or anything else that will help you to complete the job on time. Ask for help. Here is a typical prophetic conclusion:

```
                    3.0 Conclusion

Next step:  Complete criterion references, then distribute tentative
objectives to each school for criticism, allowing three weeks to
receive completed critiques.

Tentative completion:  We are at least two weeks behind schedule.
With luck we can make some of it up during the final revision stage.
Unless something unforeseen happens, we should have the course ready
to use during the first or second week of August.

Recommendations:  We could use a bit more stenographic help getting
out the tentative objectives.  Even a few hours help from a first-
rate typist would help immensely.
```

Special Features

The sample progress report used in the previous examples is, of course, quite informal. Many progress reports are short and informal, especially those sent to a supervisor or to someone with whom we work closely. Others, though, are highly formal, differing from long, formal reports only in the main text. A progress report may include a formal title page, preface, table of contents, list of illustrations, abstract, and other, less common front matter. It may also contain large amounts of back matter. In fact, the only way to distinguish such a progress report from a conventional long, formal report is by noticing the word *progress* in the title and by closely examining the main text.

The informal progress report, like most short, informal reports, has only a heading for front matter and perhaps an appendix as back matter. Here is a typical heading:

```
Service is  **         GRAY ELECTRONICS        **  8132 Seaforth
our goal         A Division of Ramco Industries      Erie, PA  61631



To:  J. C. Tremblay, Executive Vice President

Progress Report No. 3--Technician Curr. Project.

Period Covered:  May 1-30.

Submitted By:  Jorge Casiano, In-service Director
```

If your firm has guidelines that indicate the degree of formality to use, follow them. Otherwise, analyze the reader and the situation and use your judgment. The following suggestions may help:

1.   Construct a formal presentation if the report will likely be read by a large number of people in different positions or different organizations.
2.   Use length to help you decide formality. Two or three pages of main text could hardly justify complete front and back matter; twenty pages of main text needs an abstract, table of contents, and other formal elements.

3. Submit a formal progress report if copies will be used for future reference or for some other purpose such as submission to a governmental agency or a financial organization.

4. Write simple, informal progress reports rather than long, formal ones whenever possible. Be concise and direct and write directly to the reader.

5. Do not hesitate to devise your own format, using elements of both the long, formal report and the short, informal presentation. The model progress report shown in Figure 10.6 is essentially informal, but it contains several formal elements. A model letter of transmittal introducing the report to a particular reader is shown in Figure 10.7.

• EXERCISES

1. Write a formal report on local job opportunities in your major field, concentrating on the position you hope to occupy upon graduation. You determine how large a geographical area you wish to include. Your advisor or your school's placement service or career development center can help you get started.

2. Write a formal report documenting a significant project you have recently completed. Some suggestions are listed below.

    a. A major experiment in a science class.
    b. A building or design project in an electronics or engineering class.
    c. A program or system project in a data processing class.
    d. A major repair or alteration on a car or motorcycle.
    e. A major carpentry or electronics project.
    f. A sewing project.
    g. A craft project.

3. Convert the formal report you wrote earlier to a semiformal report. Reread the description of a semiformal report and study the model (Figure 10.4). Select a format and do the necessary deleting and rewriting.

4. Write a semiformal report documenting a particularly rough day at work. Aim it at your instructor, assuming that he or she has no particular knowledge of your work. If you have never been employed, document a hectic final examination week.

5. Read the local newspaper, especially the editorial page, for at least a week, and listen to as many local news, talk, and call-in radio and television shows as possible. Determine what seems to be the most prominent regional or national issue. Write a semiformal report defining the problem, summarizing major opinions, and recommending a solution.

6. Write a progress report to your instructor on your work thus far in this course. Project completion of the course, including your anticipated final grade.

7. If you are writing a formal report as a major term project, write a progress report telling your instructor how your major report is going.

8. Assume that your college education is being financed by someone—the government, an employer, or a rich uncle, for example—that requires a written progress report at about this time every semester. Write the report currently due.

F&N HEATING AND COOLING
1437 West Alping Blvd.
St. Croix, Minn.   58371

Preliminary Report #2
on
St. Croix Junior High School
Job #72A

Submitted to: St. Croix Board of Education
J. D. Willis Assoc., Architects
R. N. Delloms, City Bldg. Inspector
Joe Brinkoetter, General Contractor

Work done during: April, 1980

Submitted on: May 6, 1980

Figure 10.6 Model Progress Report

Contents

1.0   Summary to Date

Almost 23 percent of the job has been completed in 26 percent of the allotted time. The masons' strike was responsible for the delay. All equipment is in hand, and 42 percent of the duct has been fabricated. Installation of the ductwork on the east-wing first floor is complete and on the west-wing first floor is in progress. All first-floor trunklines are in place. Work in the gymnasium-auditorium complex has begun. Ductwork to begin second-floor installation is ready.

2.0   Prior Work Accomplished

Upon contract signing on March 3, we began ordering equipment and metal. During March we constructed 18 percent of the necessary ductwork. We also met with all contractors and discussed scheduling. No installation could be done, as the structural work was not ready.

3.0   Work This Period

3.1   Jobs Accomplished

1. Delivered ductwork to site--east wing.
2. Delivered first-floor equipment to storage area.
3. Began east-wing installation.
4. Continued fabrication in shop--suspending installation.
5. Recommenced east-wing installation, completing it on April 25.

6. Began west-wing installation on April 19.
7. Ran first-floor trunk lines.
8. Began gymnasium-auditorium installation on April 26.
9. Delivered ductwork to complete first floor and begin second
    floor on April 29.

3.2 Difficulties

1. Foul weather in March delayed structural work several days.
2. The eight-working-day masons' strike forced us to discon-
    tinue installation until structural work was ready.

4.0   Terminal Section

4.1 Next Month

1. Continue fabrication.
2. Complete west-wing and gymnasium-auditorium installation.
3. Install all first-floor equipment.
4. Proceed with second-floor installation as general con-
    tractors get the walls ready.

4.2 Forecast of Completion

We are only slightly behind schedule, in spite of the mason's
strike and the delayed start of structural work. Assuming no
other such unforeseen delays occur, we should be able to finish
by July 22, one week ahead of schedule. We were concerned
about possible delays getting some of the ventilating units
from Janitrol. But so far everything has arrived on schedule,
and they have assured us that there will be no other significant delays. The forecast is bright.

4.3 Recommendations

We qualify for the 18 percent payment due us at this stage. No
revisions in plans are recommended. Just keep things going as
smoothly as they are now.

F-N HEATING & COOLING

1437 ALPINE BLVD.   ST. CROIX, MINN. 58371   347-8922   347-5622

May 6, 1980

Mr. Joe Brinkoetter, General Contractor
21347 Applepall Island Blvd.
St. Croix, Minn. 58371

Dear Joe:

Here is your copy of our April report on the Junior High job. You
had us worried for a while with your masons' strike, but we managed
to keep busy.

Nothing in here is particularly important; just the usual information. We hope it will satisfy Willis's people so that the school
board will stay on schedule with the payments.

As usual, you and Bob have everything well in hand, and it looks
like everything should continue to go well.

Let's get together at Cresthaven for a round of golf one of these
Saturdays.

Cordially,

Maurice R. Fawcett

Maurice R. Fawcett
Managing Partner

MF/ss

Enclosure

Figure 10.7 Model Transmittal Letter

11

Special Applications

Overview In addition to letters, memorandums, and reports, technical people sometimes have to construct proposals, job descriptions, specifications, and journal articles.
Proposals, which require extensive preliminary planning, include both nontechnical
and highly technical sections. Job descriptions provide a basis for promotions, salary
increases, and changes in work procedure. Errors in specifications can be among the
most costly an employee can make, as they can lead to wrong or defective products or
services. Writing specifications requires great accuracy and careful checking. Technical
articles give you an opportunity to share your ideas and to gain professional recognition.
Well-constructed forms are the most efficient means of communicating routine
information.

Letters, memorandums, short reports, and long reports represent common tasks
faced by the typical technical writer. However, they by no means exhaust the
types of writing done by various professionals. In fact, it would require an entire chapter just to list and offer working definitions of all the possibilities.
Rather than do that, this chapter will concentrate on five specific types of technical writing: proposals, job descriptions, specifications, journal articles, and
forms. All are regularly constructed by large numbers of employees in many
technical areas.

Proposals

Proposals come in almost as many varieties as do reports, running the gamut
from multiple-volume, formal documents to single-page, handwritten memorandums. They may be highly technical or completely nontechnical. The
reader may be your immediate superior, your company president, a governmental agency, or a prospective client. You may propose to buy something, sell
it, build it, repair it—or you may be suggesting that someone else take one of
these actions.

Whatever the situation, all proposals entail certain common considerations.
This section will examine first what a proposal is, then how to prepare one, and
finally some of the basic varieties.

Basic Elements

Most effective proposals are organized deductively, beginning with a clear, concise statement of the basic proposal, elaborating on it in general terms, then filling in all the necessary details. Generally this is done on the two extreme levels
of technicality, explaining first in layman's terms, then elaborating in highly
technical terms. This is necessary because the person making the final decision

is often a manager or an executive who cannot understand highly technical details; yet he or she will almost always want a specialist to verify the technical accuracy and thoroughness of those details.

The basic proposal can usually be constructed in the following form: "_____ propose(s) to do _____ for the sum of _____." This statement may also include a brief indication of the manner of doing the job and the date on which it will be done. Here are some typical basic proposals:

C & E Avionics proposes to build the Duvall County Sheriff's communication system, using originally designed electronics equipment, for $27,213.

Magnavac Computers proposes to furnish and install, within thirty days of contract, a complete Magna 1670 Central Computer with five remote terminals for the sum of $82,642.

I (Elizabeth Moothart) propose to study, select, and supervise the transition to a word-processing system of our department at a cost of $1,200 plus the cost of the equipment.

Elaborate on the basic proposal by explaining more fully, but still nontechnically, what you propose to do, how and when you propose to do it, and how you arrived at the cost figure. Here is such an elaboration, based on the last of the sample basic statements shown above:

The cost would be for twenty working days' salary while I am on leave from my office-manager duties examining existing systems, evaluating hardware from various suppliers, and learning to operate and manage the system. Included in the figure is $100 to cover the cost for a one-day orientation session for our four stenographers.

My proposed plan is to spend two weeks immediately upon authorization visiting local organizations currently using such systems. Then one week would be devoted to studying and evaluating proposed systems from local suppliers. The final week would be spent during and immediately after installation in early fall. The orientation session would be held right after installation, and the system could be working by September 25.

The highly technical explanation should be complete almost to the last detail. Explain exactly how you will do any work involved. Describe completely any equipment or material furnished, including parts lists, specifications, drawings, or blueprints. Show a precise timetable and a detailed price list. The more thoroughly you can explain what, when, how, and how much, the better. Figure 11.1 shows an effective organization and tone for a short, informal proposal. The formal proposal in Figure 11.2 demonstrates a highly technical explanation.

GREEN ACRES PLAY LAND
Chippewa Falls, Wisconsin 48320

To: E. R. Bevistein, Operations Superintendent

From: P. M. Bodosky, Assistant Superintendent of Operations
           for Stock Control

Date: August 13, 1981

Subject: Proposed expanded training program for part-time
           stockroom employees

Basic proposal: I propose to run six three-hour training sessions
for all current part-time and temporary employees in my department.
Cost will be negligible, as sessions will be worked into Saturday
and Sunday night shifts when work is very slow. Any time lost will
be quickly made up because of the increased efficiency of the
employees involved. Benefits in addition to increased efficiency
include improved morale of both part- and full-time employees,
greater freedom for supervisory personnel, and improved flexibility
of job coverage.

Discussion: The first session is tentatively scheduled for Saturday, August 26, with one six-session series running on successive
Saturdays and another on successive Sundays. Exact times will
vary according to weekly work demands. Twenty-one employees will
comprise each group.

      The first two classes will be devoted to the films and
demonstrations used for training new full-time permanent employees.
Each succeeding session will include a one-hour demonstration and
two hours of guided practice on one of the four primary stock areas.
In addition, I will supervise each person's performance during the
six-week period.

      The primary benefit of the training program will be in-
creased overall efficiency of the department. This should mean a
sharp decrease in complaints from the various shop and food-
service areas. Foremen and lead workers will have more time for
their intended roles as supervisors rather than having to do the
actual stocking and order filling.

      Morale of everyone in the department as well as that of
shop and food-service managers should improve dramatically, for
reasons already stated. Also, the concerned employees will no

Figure 11.1 Model Informal Proposal

E. R. Bevistein, August 13, 1981, page 2.

longer feel helpless or inept when they are called on to perform
various functions. They will be able to adequately perform standard stockroom functions and can fill in when needed in any of
our four specialized areas. Several of them will also be able to
fill in when necessary as temporary leads.

Conclusion: I recommend that you give me immediate written authorization to inform the employees involved and to requisition
the necessary materials. Once the program is instituted, I will
give you written weekly reports.

```
                    CRANDALL BROS. BUILDERS
                    16871 West Grand Avenue
                    San Antonio, Texas   71362




                    Proposal to Build 18' x 25'
                    Screened Porch Addition




         Submitted to:  Juanita Perez
         Submitted on:  July 14, 1980
         Authorized by:  Ronald R. Crandall
         Prepared by:  Donald D. Gazdik
```

Figure 11.2 Model Formal Proposal

Basic Proposal

Crandall Bros. Builders proposes to create for Ms. Juanita
Perez an 18' x 25' screened porch to the rear left corner of her
home at 2397 Villa Puerta Way, San Antonio, for the sum of one
thousand six hundred and sixteen dollars ($1,616). All materials,
labor, and city permits are included. The job can begin within
one week of contract signing and will require eleven calendar
days to complete.

Discussion

Parts and Materials

1. 8' x 25' concrete slab attached to present slab
   4" concrete, reinforced with 1/2" steel rods
2. 16' x 25' aluminum roof attached to overhang
   16 gauge aluminum, self-braced at 1' intervals
   16 gauge aluminum gutter and downspout
3. 14-4" aluminum T beam supports--floor to ceiling
4. Green nylon screening as needed
5. 36" aluminum and nylon screen door
6. 12"-16 gauge aluminum kick panel x 61'

Other Services

The necessary building permit ($25) will be purchased by
Crandall Bros. We will also arrange the necessary inspections.

We will clean the entire area completely and will repair or
replace all damaged sod and shrubbery. Our guarantee, explained
in detail below, includes complete customer satisfaction with our
restoration of landscape. We cannot, however, transplant any
shrubbery standing where the slab will be poured.

References and Guarantees

Crandall Bros. is San Antonio's oldest builder specializing
in home improvements, now in its thirty-first year of service. We
are members in good standing of the San Antonio Chamber of Commerce
and its better business division. We are also members of the
Greater San Antonio Builders and Developers Association, and we
subscribe to its code of ethics, which Mr. Crandall helped to develop in 1969. We are happy to furnish written references and to
conduct inspection tours of our previous work. We pride ourselves
on using only the finest materials and paying premium wages to retain highly skilled employees. Our average employee has been with

us five and one-half years.

     All work is unconditionally guaranteed against defects in
workmanship or materials for one year from completion. This includes concrete, aluminum, screen, and site restoration. It does
not include damage to screening caused by negligence of the home-
owner. In addition, the aluminum roof is guaranteed against leakage for five years from installation. Should leakage occur, the
roof will be repaired free of charge and compensation will be
given for any damage caused to furnishings by such leakage.

Timetable for Construction

First day   -- Sign contract, arrange payment, apply for city
                 permit, schedule work crews.

Eighth day   -- Concrete crew reaches site, pours and finishes slab.

Tenth day   -- Aluminum crew installs roof, gutter, braces, and
                 frames screen door.

Eleventh day -- Screen crew installs screening; cleanup crew re-
                 stores site.

Conclusions

     As the details given above should indicate, Crandall Bros.
proposes to erect the finest screened porch to be built in the
San Antonio area. Our workmanship is the finest, and we use only
the best available materials. Yet our high volume enables us to
offer a quite competitive price. Consider our price, our workmanship, and our outstanding guarantee; no one can match our proposal.

Preliminary Planning

The hardest work in developing effective proposals is not the actual writing, although that is certainly important. Rather, it is the preplanning. Unless you do
the right kind, nothing you do in writing up the proposal will matter. Skill in
preplanning makes proposal writing as much an art as any specific type of technical writing and makes an effective proposal writer invaluable to the
employer.

Good preplanning involves three distinct considerations: the subject, the receiver, and the competition. When you have studied each of them, you will
know better what points to stress in your proposal.

Subject ● Begin your preliminary planning by thoroughly evaluating the
subject of the proposal. Usually the subject can best be thought of as a problem
to solve. For instance, if your subject is selling laboratory equipment, the problem is "how to equip satisfactorily the laboratory in question." If your subject is
a recommended adjustment in personnel allocations between branches, your
problem might be stated as "how better to staff each branch without new
personnel."

Now study the problem from all angles. Look first at formal printed material such as specifications, legal requirements, and fact sheets. If there are formal restrictions governing what you can propose or if the intended customer has stated specific requirements, you must, of course, work within them.

Now try to work out several possible methods of solving the problem within the formal requirements. For instance, if you were proposing to add a screened porch to someone's home, you could likely suggest several porches—one stressing economy, one appearance, and one low maintenance and long life.

Finally, double-check to make certain that you do indeed thoroughly understand the problem, that your solutions meet all requirements, and that you have not overlooked any potentially sound solutions.

Receiver ● Now shift your consideration to the receiver of the proposal. *Receiver* applies specifically to the person or group who will make the final decision about the proposal rather than to an intermediary you might actually send it to. Try to learn everything you can that will help you aim the proposal directly at the receiver. Look for likes, dislikes, quirks, eccentricities, and the like. The more you know about the receiver, the better. Discuss the problem with him or her if possible. Keep your ears and eyes open and use your good sense.

The information you gain can help you in several ways. It can tell you which of your several possible solutions to propose, and it can help you decide how to present and phrase your proposal. Assume, for instance, that you have studied the potential customer mentioned above who wishes to add a porch to her home. You have discussed possibilities with her. You have looked over not only her home but her entire neighborhood. You should now be in a good position to propose the addition most likely to meet your client's needs.

Competitors ● The word *proposal* implies competition. If there were no competition, your proposal would be a mere formality. Your competition will usually be another company or companies. Even if the proposal is merely a recommendation to your boss, you still have competition—the *status quo*. Unless you can propose to solve the problem in a way that will satisfy your receiver better than the proposals of the competition, your proposal will not be accepted. It does you no good to propose the least expensive addition to the receiver's home if one of your major competitors handles cheaper merchandise and pays his employees less. Instead you will have to stress value. Build your proposal around the notion that, although your proposed initial price may not be the lowest, it is indeed the best value in terms of upkeep and durability. If the receiver is absolutely determined to get the cheapest possible addition, you are out of luck, but otherwise you at least have a chance.

The Angle ● The example just examined shows a typical angle: value. All good proposals have some such angle. It may be low price, high quality, good appearance, almost anything. Just find some way to propose to satisfy your reader better than the competition can. Then stress this angle throughout.

One common type of proposal whose success depends almost entirely upon the angle is the proposal for a grant from a governmental agency. If you find the right angle for the particular agency at the particular time, you are almost guaranteed success. Effective proposal writers are able to determine the proper angle and to subtly imply it throughout their proposals. The results are worth the trouble, as some organizations keep coming up with lucrative grants while similar organizations seem unable to win any.

Appropriate Tone

A good proposal should exude an honest confidence. You must convince the receiver of your organizations' competence. If he or she doesn't believe you can do the job well, someone else will get it. Use two techniques to convince the receiver of your competence. First, make the technical aspects of the proposed solution accurate and thorough. Explain clearly and fully exactly what you propose to do and how you propose to do it, and make certain that the proposed plan is sound. If you are furnishing materials, make certain that they are appropriate, and describe them fully. Convince the receiver that you do understand the problem and have carefully worked out a solution.

In addition to demonstrating your technical competence, pat yourself on the back a bit. Modesty is certainly a virtue, but demonstrate it somewhere other than in your proposals. Do not simply make assertions about how good you are: give some evidence. Mention names of past satisfied clients; give references. Document the accomplishments of key members of your group. If the proposal is long and formal you can even include resumes of those key people. Convince the receiver that your organization is good.

Special Features

An exasperating but frequent problem occurs when the person who will receive your proposal gives inaccurate or inappropriate specifications or asks for something to be done in a manner you consider ineffective. Assume, for instance, that you are preparing the proposal to build the enclosed porch mentioned earlier. While studying the prospect's specifications, you notice that she has asked for a three-inch, unsupported concrete slab. Knowing the soil type and the amount of stress on such a floor, you immediately realize that at least a six-inch slab with a limerock base is needed. Such a situation is obviously very touchy. If you propose the three-inch slab, you risk incurring a customer's ire when her floor starts cracking. But if you propose something different from what the specifications call for, you will likely lose the contract, especially since your proposed floor is considerably more costly than that specified.

Either of two courses of action might resolve this dilemma. One possibility is to contact the receiver by telephone or in person and explain the situation. In an extreme case such as this—where the floor might literally cave in under someone—direct contact is probably your best course. Chances are the prospect will be pleased that you caught the error before it was acted upon. However, in less extreme instances, a wiser course might be to submit your proposal, but submit two versions. One version should follow the specifications precisely; the other incorporate the change you think necessary. Such a two-alternative proposal is often called an *exception* and is quite commonly used in cases where the specifications contain subtle errors in judgment.

When you take exception this way, meticulously justify your suggested change. Explain exactly what you are taking exception to and why you believe that your plan is superior. This is doubly important if the change would be more expensive or time-consuming, as the receiver cannot help but suspect your motives.

If you can convince the receiver that your exception is legitimate, that it is based on good motives, you are in a strong position. Should he or she want the job done as specified, you can do it knowing that you have given warning of possible troubles. Should the receiver accept your exception, he or she cannot help but appreciate your pointing it out.

Unsolicited Proposals ● Another troublesome situation occurs when you want to submit a proposal but have not been invited to do so. Such a situation can be handled in two ways. You can call or visit the prospective client and ask permission to submit a proposal. In-person contact is usually preferable to a phone call, as it gives you a chance to explain yourself more fully. The second way to handle the situation is to go ahead and submit the unsolicited proposal; but you risk having your efforts go to waste if your proposal hits the receiver's trash basket.

Should you decide to send an unsolicited proposal, include a letter of transmittal persuading the receiver to consider it. Construct such a letter very carefully. Begin by explaining how you learned of the reader's need for your services. Then explain why your proposal is likely to be superior to others he or she may receive. This is crucial, as he or she is most likely convinced that the proposals solicited will be adequate. Next, mention something specific about your organization's excellent record. A good item to include is a mention of a similar job or jobs you have accomplished satisfactorily. Use the angle you developed in your preliminary planning. Close the letter with a sentence or two expressing your appreciation for the receiver's considering your proposal and offering to discuss the proposal at his or her convenience. Such a letter, Figure 11.3, follows the proposal in Figure 11.2.

Job Descriptions

Technical and professional employees with supervisory responsibility are sometimes called upon to prepare job descriptions of their own positions and positions under their supervision. Should you face such a task, you will probably find it difficult. While you might describe in one paragraph "off the top of your

Crandall Bros. Builders
16871 West Grand Avenue
San Antonio, Texas   71362

July 14, 1980

Ms. Juanita Perez
2397 Villa Puerta Way
San Antonio, Texas   71355

Dear Ms. Perez:

Enclosed is our proposal to erect your screened porch. You should find our proposal competitively priced at $1,616 for the finest porch you could have built by anyone in the metropolitan area. We use the best materials and have the best-qualified staff available; but our strongest feature is the only unconditional guarantee offered by any South Texas builder.

If you have any questions, just let me know and I will be glad to meet with you at your convenience. We are certain you would have many years of enjoyable outdoor living from your Crandall Bros. porch, and are anxious to begin work.

Cordially,

Donald D. Gazdik
Home Improvement Supervisor

DG:ln

Enclosure

Figure 11.3 Model Transmittal Letter

head" your job and each position under your supervision, odds are great that these descriptions would be practically worthless. Good job descriptions require careful analysis and equally careful writing.

Preliminary Analysis

Before you can write a job description, you will have to analyze the job; and to do that correctly, you must establish what you hope the description will accomplish. For what purpose will the results be used? Some common functions are listed below:

1. To ascertain job requirements in order to facilitate placement, hiring, promotion, and transfer of employees.
2. To standardize jobs, clarifying what each person is expected to do, thus simplifying both his or her work and that of the supervisor.
3. To isolate specific tasks performed so that they can be studied to increase efficiency.
4. To aid in more effective recruitment by the personnel department.

The actual analysis consists of two basic phases—gathering the information and analyzing it. Information gathering, obviously the first phase, can be accomplished by any of three basic methods: questionnaires, interviews, or direct observation. If, as will usually be true, you are dealing with only a few different jobs, a combination of observation and interview will probably work well. If you wish to use more formal techniques, check any good textbook in personnel management.

If you interview people as part of your analysis, be certain to do some advance preparation. Casual chats are great, but take along a list of specific questions to ask or items to discuss. You can phrase your own questions to fit the specific jobs in your organization, but the list below will suggest kinds of information you should seek:

1. Description of basic duties performed.
2. Experience needed to qualify for the job.
3. Break-in time needed to become proficient.
4. Special knowledge needed for the job.
5. Routine tasks performed daily.
6. Regular tasks performed less often.
7. Occasional tasks.
8. Special duties.
9. Supervisory responsibilities.
10. Unusual or unpleasant working conditions.

After you have talked with enough people and are satisfied with your observation, study your results. Your major concern will be to reconcile the responses from different employees in the same position. Keep in mind that you are analyzing a job, a position, not the person in it. Remember also that you are concerned with what the job is, not with what it should be.

Basic Elements

The first step in composing the job description is to make certain that the job title is appropriate. Check the *Dictionary of Occupational Titles,* a document published by the U.S. Department of Labor giving titles and brief descriptions of some 40,000 positions. *Occupational Guides,* a set of pamphlets also published by the Department of Labor, might help as well.

The actual description may vary from one concise, well-developed paragraph to several pages. The single-paragraph version is essentially a summary of the longer, more complete version. The thorough description comes in as many forms as there are organizations that use it. Use your good judgment based on what you think the situation demands. Following is a list of information you might include:

1. Name of job (preferably from *Dictionary of Occupational Titles*).
2. Description of duties performed.
3. Materials and equipment used.
4. Special conditions needed, such as lighting, ventilation, and noise control.
5. Relationships to other positions:
   Departments work sent to or received from.
   Supervisory responsibilities and source of supervision.
   Promotional opportunities.
6. Prerequisite knowledge.
7. Special qualifications:
   Experience.
   Physical or mental tests.
   Degrees.
   Special certification or registry.

Figure 11.4 shows a model job description.

Specifications

The challenge of writing specifications lies in the absolute precision and clarity required for the task. Every word, every comma, in a specification is crucial. Stories abound about misplaced commas in specifications costing thousands of dollars or causing multimillion-dollar lawsuits. Should you ever have to write a specification on the job, strive for precision from the start of your preliminary planning through the final proofreading of your final draft.

```
                           JOB DESCRIPTION


Department   Nursing Service

Job Title    Ward Clerk

Date         4-4-80

Primary function:  Perform all general clerical and receptionist
duties for department.

Tools and equipment used:  Addressograph, manuals and medical
reference books, patient charts, doctors' orders, copy machines,
filing equipment, Kardex Plus cards, numerous printed forms,
telephone, typewriter.

Supervisor:  Head nurse of ward.

Supervision exercised:  None.

Major jobs performed:  (These are the principal functions only;
not all tasks required are listed.)
 1.  Keep all ward records including but not limited to the fol-
     lowing:  unit census, admissions, transfers, discharges,
     check-in and check-out sheets, food tray checklist, patient
     records and charts, narcotic lists, doctors' orders, Kardex,
     unit correspondence.
 2.  Maintain supplies and cleanliness in nursing station.
 3.  Receive, route, and page all telephone calls.
 4.  Receive and deal with all visitors.
 5.  Requisition services from dietary, laboratory, pharmacy,
     housekeeping, laundry, maintenance, central supply, and
     storeroom departments.
 6.  Distribute materials within department and run errands.
 7.  Receive and handle all paperwork for each patient.
 8.  Handle all notifications and paperwork in case of patient
     death.
 9.  Conduct fire inspections and drills.
10.  Maintain adequate stocks of all supplies.

Qualifications:  High-school diploma, one year's training in voca-
tional school or nursing school, standard physical examination,
one semester typing, three-month internship with full pay.

Salary:  $8,500-$9,500.

Promotional opportunities:  Chief Ward Clerk if opening becomes
available.
```

Figure 11.4 Model Job Description

A specification is a detailed description of a piece of merchandise or a service you are planning to purchase. It is your specification that a proposal writer uses in preparing the proposal. You must stipulate exactly what services or materials you expect to receive. The organization whose proposal you accept will perform services and provide materials exactly as specified. Obviously, if your specifications are inaccurate, vague, or ambiguous, you can expect serious problems. Once you contract with someone according to certain specifications, you can require that supplier to comply only with what the specifications actually say, not with what you might wish them to say or might have originally intended them to say. In fact, most court cases concerning interpretations of specifications are decided in favor of the supplier.

Structure

Specifications have much the same basic structure as proposals: they first explain what you want in general terms, usually in nontechnical language; then they explain each aspect of the service or material being specified in thorough, highly technical terms. Assume, for instance, that you are preparing specifications for a small building to use as a branch office. First explain generally what you want, covering such items as dimensions, type of materials, kind of roof, type of heating and cooling plants, and type of interior finish. Specify the date by which you want the building completed, the method of payment you propose, the dates by which you want proposals submitted, and the date on which you will award a contract. Mention also any special requirements, such as a performance bond or proof of competence by the builder.

Next write an individual section for each technical aspect of the project (often called technical clauses). Typical sections might be grading, excavation, concrete, framing, plumbing, electrical, flooring, finish carpentry, roofing, and landscape. A final section assigns responsibility for making changes in the specifications.

Each technical clause should contain precise details of what is wanted. Quality standards must be included, in precise terms. Do not use vague statements such as "highest quality" or "satisfactory to buyer" or "of appropriate thickness." State exactly what you want and how you will determine quality. Refer to industrial standards when appropriate. Give exact weights, thicknesses, and grades of material. Remember that the average contractor (contractor is used here in a broad sense to indicate the person accepting a contract to furnish goods or services) is not as concerned about quality as is the customer—you, in this case. He or she often wants only to give quality high enough to meet your specifications and to protect his or her reputation.

Begin each clause with an introductory section describing the activity or material included and relating the clause to the overall project. Next write a body with one or more paragraphs for each part of the clause, covering materials, workmanship, and methods. Be thorough and precise. Include drawings wherever appropriate. Complete the clause with a brief terminal section, usually one or two paragraphs, in which you state how you will determine satisfactory completion of the work entailed in this clause and how you will pay for it.

General Guidelines

The following guidelines will help you with the careful planning, constructing, and editing required for accurate, successful specifications.

1. Unless you are completely authoritative on every aspect of the project, get assistance in preparing specifications. A specialist in each field covered should develop the appropriate technical clause. If that necessitates paying outside consultants, do so. Each clause must be technically accurate and thorough. You can then develop the general section and a table of contents and can edit the technical clauses to get them into similar form and style.

2. Use the simplest, most direct writing style possible. Avoid legal jargon or technical jargon. Clear, direct language with short, simple sentences will be least likely to cause future complications.

3. Use standard symbols and abbreviations, especially in the technical clauses.

4. Cross-reference where appropriate to avoid repeating information given in another clause.

5. Refer to published, accepted standards of professional groups if the standards are well known in the field. Feel free also to refer to well-known, accepted methods or procedures. Catalog and manufacturers' literature can also be cited, but avoid confining a would-be contractor with too many brand names.

6. Use illustrations liberally, but be sure they are accurate, thorough, and understandable.

7. Make the general section nontechnical enough to be understood by a manager or other executive, but make all other clauses highly technical.

8. Include in the general section a set of working definitions, if necessary, but try to avoid using terms requiring definition.

9. If you think you may want to make any changes after the contract is awarded, include in the terminal section of the appropriate clauses a concise statement giving you the right to request changes and giving the contractor the right to fair compensation.

10. In writing clauses, specify what you want precisely enough to assure quality; but try to allow the contractor some flexibility to use innovative methods and new materials.

11. Specify a desired completion time. This may be done by either stipulating a specific date or a certain number of calendar or working days.

12. Use *shall* when writing about the contractor, *will* when writing about the owner (you).

13. Remember—be precise, be thorough, and proofread carefully.

Many large companies and other organizations publish short articles by employees who write about new equipment, techniques, and approaches.

Professional Articles

Technical or professional persons with good writing skills not only write more effectively on the job than their colleagues; they are also able to enhance their professional status (and sometimes increase their bank balances) by contributing articles to journals. Getting such articles published is not nearly so difficult as many people assume. If you have technical competence, solid writing skills, and the energy to do a little more preliminary planning than you would do for most pieces of writing, you can be published.

Selecting a Journal

Technical periodicals in which you might publish articles can be classified into three distinct groups: in-house organs, trade and popular publications, and professional journals. The in-house publication is a good place to begin and to refine your techniques. Most large companies and other organizations publish various sorts of newsletters and periodic bulletins. Designed essentially to improve intra-organizational communications and to build esprit de corps, such publications regularly seek short articles by employees telling about new equipment or new techniques and approaches. If your organization distributes such a publication, check with its editor.

Trade and popular journals can be found for almost any occupational area—usually several per field. Such journals serve several functions: They keep members of the field apprised of new developments in technique and equipment through both articles and advertisements. They act as clearinghouses for tips about tricks of the trade. They help managers and laymen interested in the field to keep abreast of basic trends. Nearly every skilled, experienced professional has something to contribute to such journals. The journals often compete actively for good articles, paying off not only in prestige and personal satisfaction but also in fees.

Professional journals are probably the most desirable place to publish. Although they rarely pay cash for articles, they pay off handsomely in prestige. Nearly every branch of every major profession has a least one such journal. In these journals are printed the best current highly technical articles available. Through such journals many advancements in science and technology are transmitted, experts exchange information, and beginners supplement their college training with the latest knowledge. Publication in such a journal will almost certainly enhance your standing within your organization, and it will help you to build a reputation in the field, even on a national or international level.

Writing and Submitting

Do not commit yourself to the serious, time-consuming task of writing the article until you are certain of two matters: that the article will be given serious consideration when received and that you are submitting it in the proper form. If you are uncertain about either the editorial policy of the journal or its style and form requirements, write the editor. Some experienced article writers recommend that you send such a letter to any editor who is new to you. One way or another, be certain that the editor will consider publishing your article and that you know the appropriate manner of submission.

Composing the article is much like composing any other written communication. Answer the standard preliminary qustions; build an outline; prepare a rough draft; revise and polish it. Do up a final copy or copies. Exercise the same care you would with the most critical report or proposal.

Although your manuscript should be prepared and submitted in exact accordance with the wishes of the editor, the suggestions below can help:

1. Most major professional organizations endorse or establish certain style manuals or publication guidelines. Check with yours.
2. Handle abbreviations, numbers, illustrations, references, and subheadings according to the editor's wishes or the appropriate guidelines or style manual.
3. Type your manuscript on good bond paper (Sub. 20 is preferred). Use pica type if possible. Submit one extra copy unless more are requested. Keep a copy for your own reference.
4. Double-space and leave ample margins.

5.   Put your name, title, and business address in the upper right-hand corner of the first page.

6.   Submit illustrations separately from the typed text, but clearly indicate their placement.

7.   Mail the manuscript flat, using a heavy piece of cardboard to protect illustrations.

8.   Include a self-addressed, stamped envelope in case the manuscript is rejected or must be revised.

Follow-Up

High-quality journals have exacting editorial standards. Many use editorial boards to evaluate potential manuscripts, so you might not hear of your submission's fate for some time. However, if six weeks pass without word, you can send a follow-up inquiry.

Should your manuscript be rejected, most editors will tell you why. The article could be too long or too short, in which case you can revise and resubmit. The subject may not be appropriate, in which case you can submit the manuscript elsewhere. Often an article is rejected because a journal has a policy against publishing too many articles on one subject, so again you can submit elsewhere.

If your article is rejected for lack of quality, do not give up. Study the editor's criticisms. If the weakness is stylistic, brush up on your technique. If the weakness is a matter of content, do more research or find a more salable subject. Remember, all writers get rejections.

Forms

Of all types of written technical communication, perhaps the most inescapable are forms. Nearly every professional or technical person has at least occasionally to complete forms, many of us must do so regularly, and others are inundated with required forms. In some organizations, forms are required for nearly everything that goes on, first to secure permission to do it, then to verify that it was done. In a few extreme cases, one form must be completed just to gain authorization to complete a larger, more important form.

Advantages and Disadvantages of Using Forms

For most routine written communication, forms are clearly the most effective means to use—good forms, that is. Poorly constructed ones negate the advantages of using forms and introduce many disadvantages. The following list presents the advantages of using forms for ordinary, day-to-day communication.

1.   They can be completed quickly. A well-designed form of even two or three full pages can be filled in in ten or fifteen minutes.
2.   They assure completeness of information. If you fill in forms carefully, you run little risk of omitting important information, and you do not have to wrack your brain to be sure that you are including everything.
3.   They limit inclusion of irrelevant or trivial information. If you supply just what is called for and add additional comments only when absolutely necessary, you run little risk of putting in unneeded information.
4.   They ensure logical organization. If the form is designed to put information in a logical sequence, you are spared the trouble of devising such a sequence and the reader is guaranteed one.
5.   They ensure uniformity. The more a certain type of information is transmitted, and the more people transmitting it, the more useful forms become. A district manager receiving weekly or monthly reports from six or eight local managers or an administrator receiving reports from a dozen department heads almost has to use forms to get reasonable uniformity.
6.   They facilitate processing by electronic equipment. If information will be fed into a computer or processed through other office machines, forms can be designed that fit into the machine or at least convey data in the form needed. This eliminates one step in handling the data, thus saving both time and money.

The following list gives some disadvantages of forms. Notice that most are problems only with poorly designed forms.

1.   They are difficult to design and prepare. The less a form will be used, the less its advantage in saving time over some other means of communication. Although you might need only a few minutes to prepare a simple form for a common, short report, to do a good job requires much longer. There is no formula for determining how often a form will have to be used to make it worth the preparation time, but you certainly would not want to prepare a form for something you anticipated writing only once or twice.
2.   They can make it difficult to give certain types of information. Preparing a form is much like preparing a set of questions. You control the information the users give by asking only certain questions; and, should you not ask enough questions or ask the wrong ones, users will have difficulty saying what they want to. Suggestions for avoiding this problem are given in a later section.
3.   They can be difficult to fill in. This is primarily a problem with poorly constructed forms. It includes several different weaknesses. "Too little space" is a common complaint when a question requires an answer longer than the space provided. "I cannot understand the instructions,"

and "This question doesn't make any sense" are equally common complaints. "Why do they need to know this?" and "How am I supposed to
know that?" are questions frequently asked by users of poorly constructed
forms. Figure 11.5 shows a widely circulated example of a poorly constructed form.

COMPLAINT BLANK

State Nature of Complaint
in This Space

WRITE LEGIBLY!
GIVE FULL DETAILS!

Figure 11.5 A Poorly Constructed Form

Constructing Effective Forms

A good form must get the maximum information with the least effort on the
part of the writer. This implies two areas of concern: getting the information
you want and making the user's job as easy as possible. To these we should add
two more considerations: simplifying the handling, processing, and filing of the
completed forms and reducing printing and distribution costs.

Before you begin designing and constructing a form you must answer certain
basic questions:

1.   Is it necessary to record this information?
2.   Can some existing form be modified to include the information?
3.   Will the form be used enough to justify the time required to
develop it?

Once you have clearly established that a new form is needed, you can begin
your preliminary planning much as you would for any other piece of written
communication. Make certain that in making this preliminary study you discuss the proposed form with a good sampling of all those who will be involved:
those who will fill it in, those who will receive and read it, and those who will
process or file it. Study the form or forms it is replacing or the memorandums or
other documents conveying the same general information. You might also want
to check similar forms used by other organizations.

Now you are ready to begin building a list of questions. Using a rough outline of the material to be covered, develop sets of questions for each major topic or subtopic involved. The following guidelines will help in developing questions:

1. Design questions that require as little writing as possible. The best questions are those that can be answered with a checkmark, a circle, or an underline; next are those requiring a number, a word, or a brief phrase. Only as a last resort should you require essay-type answers.

2. Look first for either-or questions. The choices may be true-false, yes-no, completed-uncompleted, budgeted-unbudgeted, or any similar pair. Then you can print the two answers beside each question for the reader to circle, underline, or check.

3. If two alternatives will not work, try three. Make *One* positive, *two* neutral, and *three* negative. If you want better discrimination, five alternatives should work; *one* and *five* can be very positive and very negative; *two* and *four* mildly positive and negative; and *three* normal or neutral. Do not use more than five choices, as discrimination among alternatives becomes more like guesswork with each additional possibility. Nor should you use an even number of choices. Always have a middle or netural choice and the same number of negative and positive choices. Clearly indicate which is which. Do not make the user guess whether *one* or *five* indicates "excellent."

4. Use checklists. Give users a number of alternative responses and instruct them to check the most appropriate. Always include "other," with a blank for users to fill in a response not listed. If you do not give users this alternative, you are saying, in effect, that no one could possibly want to indicate anything that you could not anticipate.

5. Let the reader fill in blanks with short answers. Write out sentences with one or very few words missing. Furnish the units in which the answer is to be expressed; give all needed formulas and equations, leaving only the unknown portions blank. In short, require as little writing and computing or determining as you can.

6. If you must ask the user to write out sentences or longer passages, spell out exactly what you want. Give thorough guidelines. Never merely say "discuss" or "explain."

7. Double-check each question to make certain that it is as clear as you can make it and that it does actually ask for the information you want. Try your questions out on one of the people who will be using them or on someone of comparable background.

8. Now check the coverage of your questions. Do you have a question for every piece of information you need? Are there questions for everything a user of the form would need to convey? Is there adequate allowance for comments or "other" responses?

Now you are ready to construct the form. Begin by constructing the introductory or identification zone (sections of a form are generally called *zones*).

This always comes first. It should include the following:

1. Name of the company or organization.
2. Space for indicating the appropriate branch, department, or division.
3. Title of the report or form.
4. Space for indicating the date of submission.
5. Space for file or reference numbers.

Next, design the main body one zone at a time. The organizational pattern of the zones will vary just as the organizational pattern of any report or other written document will. If you have organized your questions in a logical sequence, developing zones should be a simple matter of dividing the questions into logical groups. Reviewing classification (Chapter 3) might help you here.

Only one section of the form remains—the conclusion. In fact, many organizations do not even have a concluding section, letting the final zone of the main body conclude the form. Should you decide to use a separate conclusion, include the following:

1. Space for the user's signature (with title) and for the date signed.
2. Space for any needed approval signatures, titles, or dates.
3. Identification number and title. Since many forms are carried in clipboards or bound at the top, putting this information at the bottom allows for quicker recognition.

Your final step is to check your nearly completed form for effective design. Use the following checklist:

1. Are there writing lines to use in answering questions? Lines are much easier to use than blank spaces. Make sure that there are plenty of lines for a complete response to each question. If the users will usually type the form, make the writing lines fit typewriter spacing. If the form will usually be completed longhand, allow at least one and one-half times as much space between lines and at least one and one-half times the number of lines needed for a typewritten answer.
2. Are the instructions clear? Keep general instructions at the beginning of the form to a minimum. Do not expect your reader to read, understand, and remember a long list of preliminary instructions. Rely mainly on instructions for each zone or question.
3. Have you furnished as much information and required the user to write as little as you can? Have you included all units?
4. Are zones set up so that the form can be completed sequentially with no skipping ahead or backtracking?
5. Is the form an appropriate size for handling and filing?
6. Have you used color to help identify similar forms and copies of a form going to different people or offices?
7. Have you used carbon paper or self-copying paper for ease in completing multiple copies?

Figures 11.6–11.8 show well-constructed forms.

MEMORIAL HOSPITAL   XENIA, OHIO
INCIDENT TO PATIENT

Incident to
Mr. Mrs. Miss_____Age____
                (Last Name)   (First Name)   (Initial)

_____
    (No. Street)   (City)   (State)   (Zip)
Room no._____ Bed no._____
Present diagnosis_____

Condition before incident ○ Normal   ○ Senile   ○ Disoriented   ○ Sedated
                           ○ Drug
                           Time of Adm._____ am
                                              pm
Exact location of incident _____Date_____Time_____ am
                                                                            pm

- -

Type of incident:
 ○ 1 Fall from bed   ○ 1 None apparent
 ○ 2 Fall while ambulatory   ○ 2 Burn
 ○ 3 Other accident in bed   ○ 3 Internal
 ○ 4 Unauthorized out of bed   ○ 4 Fracture(s)
 ○ 5 Other accident while ambulatory ○ 5 Sprain
 ○ 6 Surgery accident   ○ 6 Hematoma
 ○ 7 Missing sponge   ○ 7 Abrasion
 ○ 8 Medication error   ○ 8 Laceration
 ○ 9 Treatment error (complete material attached)
 ○10 Other--specify_____

Description of incident_____
_____

Date of report_____Signature of person reporting_____
Physician notified_____Date_____Hour_____ am
                                                                       pm

- -

Suggestions to prevent recurrence_____
_____

- -

            First employees at site   Other patients in room
Name_____Position_____ Name_____Position_____
_____   _____
                                     _____

Physician's statement of injury and trt._____

Date seen_____Time seen_____ am _____M.D.
                                       pm

- -

Routing: original to Executive Offices, copy to Dept. Head

Dept. Head signature_____ V. Pres. signature _____

Figure 11.6 Sample Form 1

POOLS
by Max

Swimming Pool
Construction & Service

SWIMMING POOL CONTRACT   195 Highway 17-92–Longwood, Florida 32750

Pools by Max, Inc., agrees to build a swimming pool according to plan as accepted and described herein for:

Name: _____

Address: _____

Location: _____

Legal Description: _____

Access: _____

Pool Shape: _____

Greatest	Shallowest	Greatest	Greatest
Depth _____	Depth _____	Length _____	Width _____

Type of Tile: _____ Type of Coping: _____

Marbelite interior finish, concrete steps with black tile edges. Electrical includes: 400 watt light, light switch and timer.

Filter System: Shall consist of skimmer, sand type fiber glass tank, motor and pump, automatic chemical feeder, strainer, gauges, vacuum connection, piping, valves and other equipment as required form complete filter system.

Filter Site: As per plan.   Cool Deck: As per plan.

 Accessories and Equipment: Vacuum, hose, vacuum head, 8' to 16' pole, 18" brush, leaf skimmer, test kit, and grab rails with inset steps.

Additional Accessories: Diving board _____ Slide _____

_____

Other Work: _____

_____

Elevation: _____

Notice to the Buyer: (a) Do not sign this before you read it or if it contains any blank spaces. (b) You are entitled to an exact copy of the paper you sign.

CONTRACT PRICE is _____ Dollars $ _____

Payable 10% herewith as a deposit; 50% upon placing of pool shell; 35% when pool is ready for pool cote and 5% balance upon completion and pool placed in operation.

Customer accepts pool as complete when he first uses it as a swimming pool or permits such use.

Upon signing, owner acknowledges his or her understanding of contract and ural structural guarantee, general terms, and conditions on reverse side.

Executed in duplicate, copy of which was delivered to, and receipt is hereby acknowledged by, buyer, this _____ day of _____ 19 _____.

Accepted _____ (Owner) Owner's Phone _____

_____ (Owner) Date Signed _____

Submitted by _____ Approved by _____

Figure 11.7 Sample Form 2

DRILLING DEPARTMENT--SUBSOIL INVESTIGATION

Willkerson Associates

Date Submitted_____Dates of Investigation_____to_____

Submitted by_____Person(s) Involved_____

Location of Investigation_____

==

Holes Drilled

#1_____#2_____#3_____
       Depth   Depth   Depth

#4_____#5_____#6_____
       Depth   Depth   Depth

Attach area map locating position of each hole

==

Preliminary Evaluation of Site_____

_____

_____

Number of Samples Taken_____Sent to_____on_____
                                                                   Date

Comments on Procedure, Equipment Condition, Personnel, Other_____

_____

_____

==

Signed_____Date_____

Attach Form #163 Travel Voucher

         #117 Location Map

Wilkerson Associates #162 Subsoil Investigation

Figure 11.8 Sample Form 3

• EXERCISES

1. Write a proposal to your instructor suggesting a topic for your major report for this course. Include a brief rationale explaining why you chose the subject and why you think it will lead to a good report. Summarize briefly the main points you intend to cover.

2. Write a proposal to the appropriate dean presenting a solution to some campus problem—parking, registration, or cafeteria, for example. Be sure to state the problem clearly before presenting your solution.

3. Assume that you have just come up with an idea for a miraculous energy-saving device—carburetor, solar panel, insulating technique, or the like. Write a proposal convincing a local bank to lend you $10,000 to build and test a prototype.

4. Write a job description for your present or most recent job.

5. Consider your present role of student as a job, and write a job description for it.

6. Write a job description for the job you would like to get upon graduation.

7. Use the guidelines on pages 250–251 to analyze a form. Select a common form from your employer or perhaps your college—accident report forms, travel vouchers, and application blanks are good ones to work with. Write a one- to two-page memorandum presenting your analysis.

8. Assume you and several other people are going to have to regularly submit reports similar to one of the memo or semiformal reports you wrote earlier in the term. Prepare a form on which reports can be submitted.

12

Individual Oral Presentations

Preparation

> Analyzing the Problem
>
> Analyzing Your Audience
>
> Building the Informative Presentation
>
> A Final Checklist

Delivery

> Choosing a Method of Delivery
>
> Using Your Voice Effectively
>
> Using Your Body Effectively
>
> Using Words Effectively
>
> Using Visual Aids

Some Special Situations

> The Persuasive Speech
>
> The Goodwill Speech

Exercises

Overview   *To prepare an effective oral presentation, analyze your audience and situation carefully. Make certain that your organization, introduction, and conclusion are tight. In delivering the speech, whether you read it or speak from notes, try to use varied tones of voice, gestures, and facial expressions that fit your topic. Use appropriate visual aids. And practice: it will help you to feel more secure and appear more natural.*

Technical writing requires much hard thought and effort, but it seldom calls for anything like the sweat and worry that go into an individual oral presentation. Giving speeches is traumatic, not just for beginners but even for veteran speakers. Stage fright is almost always there; so do not expect any magic cure for it from this chapter. It can show you how to minimize stage fright; the rest is a matter of experience.

Careful preparation will always contribute to the success of your oral presentations. If you do not prepare properly, no amount of skill can make your presentation entirely successful. On the other hand, good preparation can lead to an effective presentation even if your speaking techniques are weak. For that reason, this chapter is devoted about equally to methods of preparation and methods of delivery.

Preparation

Preparation for a speech includes three distinct phases. First analyze the problem; next analyze your audience; then build the speech. Each phase involves several distinct steps.

Analyzing the Problem

The preliminary questions to answer before preparing an oral presentation are the same ones discussed in Part I. Here we will focus specifically on the aspects of preliminary planning that most apply to oral presentations.

The question "What is it?" will usually be at least partially answered: It will be some sort of oral presentation. Now make your answer more precise. Is it an oral report? A casual chat before a small group? A formal speech before a large group? Usually the person who asks or tells you to make the presentation will state exactly what is required. If not, determine just what is expected of you.

Now choose a subject. Often you will be given a specific subject—as in the case of an oral report—but other times only a broad, vague subject will be suggested, and sometimes the choice will be entirely yours. If you do have some freedom of choice, select carefully. Use these four basic guidelines:

1.   Select a subject that interests you. If you are not interested, no one else will be.

2. Select a subject that you know something about. The more you know about it the better. If you will need more information, pick a subject about which you can easily find such information. One of the most serious mistakes made by inexperienced speakers is trying to speak on subjects they know little about. Even if you do have enough information to get by, your lack of thorough understanding will generally show through.

3. Pick a subject that fits your audience. Find something that interests them, and be sure that it is neither too technical nor too elementary.

4. Choose a subject that you can handle adequately in the time available. Some topics that make excellent thirty-minute speeches simply cannot be done justice in ten minutes; conversely a snappy, meaty ten-minute topic dragged out over thirty minutes will put your audience to sleep.

Armed with a good subject, you can next determine a purpose for your presentation. This involves two steps: First select one of the three basic purposes—to entertain, to inform, to persuade—then make that purpose more definite in terms of your chosen subject. After examining the three basic purposes, you can see how to develop a definite statement of purpose.

To Entertain ● Your sole concern here is that the audience enjoy themselves. If they have a good time, your presentation is a success; if they sit glumly or are obviously fighting to stay awake, it is a failure. Such speeches are unquestionably the most difficult for most of us, so we are fortunate that they are also the least common. Even more fortunately, such speeches are usually given voluntarily, and we need not volunteer unless we are confident we can succeed. The most common situation requiring a speech to entertain is after dinner at a social function, a special banquet, or the meeting of a club or fraternal organization.

Should you accept an invitation to make such a speech, keep two points in mind. First, remember that we are not all comedians. Few of us can tell a steady stream of jokes, keeping the audience laughing throughout. A few well-chosen stories or personal anecdotes, with perhaps two or three well-placed jokes, will generally go over better than a series of one-liners. Second, remember that your purpose is to entertain, and do not mix in anything serious. A bit of light entertainment will improve nearly any serious speech, but the reverse is not true.

To Inform ● When your general purpose is to inform, you will be presenting new ideas or trying to help your audience understand old ones. Most of what a teacher says to students is intended to inform, as is an introductory lecture given new employees, the demonstration of a new technique or piece of equipment, and the oral presentation of a report.

In each of these instances, the goal is to make the audience understand something, so the speaker must do more than merely prepare a large amount of information to present. He or she must carefully analyze the audience to determine their level of understanding and build the presentation from that level. Each step must be carefully built to fit with the previous one, so that the listeners are led step by step from the familiar to the unfamiliar. Each new point must be supported with enough detail that the listeners can incorporate it into what they already know.

Most of the oral presentations you are likely to make in your professional life fit into this general category, so most of this chapter is aimed specifically at the preparation and presentation of informative speeches.

To Persuade ● When you attempt to change your listeners' beliefs or attitudes or spur them into a particular action, your general purpose is to persuade. Most professionals occasionally have to try to persuade. A few common examples are the sales talk, the pep talk, and the attempt to gain approval for a new idea, technique, or other proposal.

As with any other type of presentation, analyze your audience carefully. If you want them to end up thinking a certain way, you must determine how they think to begin with. Rely primarily on logical arguments supported by facts and examples, making certain that these facts and examples are aimed directly at the interests of your audience.

Once you have established the general type of presentation, you can establish a specific purpose. Decide exactly what you wish the audience to think, feel, or do as a result of listening to you; then prepare a statement of purpose designed to bring that about. The following examples demonstrate three specific purpose statements based upon general purposes and subjects.

Subject: The lighter side of the emergency room.
General purpose: To entertain.
Specific purpose: To show members of the high-school paramedical club that things are not always so grim in the emergency room as most people believe.

Subject: Duties of an emergency medical technician.
General purpose: To inform.
Specific purpose: To show the high-school paramedical club exactly what an EMT does in real life as contrasted with the image presented on popular television programs.

Subject: The rewards of operating a small business.
General purpose: To persuade.
Specific purpose: To convince possible franchise buyers that owning their own franchise will give them both monetary and psychological rewards.

With a specific purpose clearly determined you are ready for the two remaining phases of preparing your presentation: analyzing your audience and building an outline.

Analyzing Your Audience

Communicating directly with the receiver of your communication may be even more important when speaking than when writing. Many oral presentations that could have been effective were not, precisely because they were not accurately directed at the audience.

Begin your audience analysis by determining some basic facts about them. Find out the approximate size of the group. Determine whether the group will be homogeneous (all pretty much alike) or heterogeneous (widely diverse). Determine the general age and sex makeup of the group. Learn their occupational interests. Determine their membership in social, professional, and religious groups. Finally, check their cultural, political, and ethnic background. If the group is relatively homogeneous you can sometimes develop a profile of the typical member. Even if the group is quite heterogeneous you can find one or two points of commonality. After all, something is bringing them together to listen to you.

Now that you know something about your audience, you can analyze them more closely. Begin with their level of technicality, their knowledge of your broad subject. If several levels are apparent in the audience, you will have to speak so that the lower level can understand yet try not to bore the more knowledgeable.

Next consider the audience's attitudes and beliefs. Three attitudes deserve your attention: attitude toward certain basic values; attitude toward your specific subject; and attitude toward you, the speaker. Determine whether your listeners share any predominant ethical, political, or philosophical attitudes. Are they all ultraconservatives, young radicals, religious fundamentalists? You cannot avoid offending someone sometime; but you can anticipate attitudes or beliefs that can swing an entire audience for or against you.

Next focus on probable attitudes toward your specific subject. Is the audience interested in your subject already, or will you have to build an interest? Will they likely hold preconceptions that you will have to overcome or that you can build on? If your general purpose is to persuade, you must determine whether your audience is hostile, neutral, or friendly.

The most difficult attitude to determine, but the one that will probably affect audience reaction most strongly, is their attitude toward you. Do they know you? In person, or by reputation? Is what they do know about you likely to be regarded as positive? If they have no prior knowledge about you, are they likely to react strongly to you—your voice? manner? appearance? To gain a favorable reaction to your presentation you will need your audience's respect, and if at all possible their admiration. If they do not like you and do not trust you, your job will be nearly impossible.

Closely related to audience analysis is your analysis of the speaking situation. Common sense says that the same group will react differently in different situations, so you really cannot be confident of your audience analysis until you have also taken the situation into account.

Is the audience volunteer or captive? Voluntary attendance indicates interest in you, your subject, or both. Captive audiences, there because they have to be, are not necessarily openly hostile, but they certainly require careful handling. What is the purpose of the occasion? Is your presentation subject geared directly toward this purpose? Are you the only speaker, or one of several? Are the other speakers' subjects related to yours? Do the physical aspects of the speaking site present any problems? Poor acoustics? Microphone? Will drinks or a meal precede your speech? Answer these and any other questions you think of about the occasion, then consider your answers together with your knowledge of the audience. Do not let yourself be surprised by having to speak into a microphone for the first time. Do not plan a straightforward, perhaps almost dull presentation to be given after the audience has eaten a full meal, had a few drinks, and heard four other speakers.

As you gain speaking experience you may be able to do much of this analysis in your head and can probably just store your results there, but for the time being you will be safer if you actually jot down an outline-form analysis to refer to in planning your presentation. Below is such an outline.

1.   Specific purpose: To explain to members of the Downtown Sertoma Club exactly what hypnotism is and to explode some common myths about it.
2.   Audience: Downtown Sertoma Club of Gotham City. Usual attendance is forty-five, all males, ages 30–65.
Commonality: Community service orientation. All professionals who also make contacts as businessmen, attorneys, CPAs, and so on. Somewhat religious and politically conservative. Mostly college graduates. No ethnic considerations.
3.   Knowledge of hypnotism: No practitioners, few if any subjects. All probably therefore nontechnical.
4.   Attitudes: General conservatism, but belief in progress, science, and so on. Likely skepticism about serious value of hypnotism. Probable belief that it is hocus-pocus; some possibly see it as dangerous. General acceptance of me as trained, certified professional. No likely strong personal reaction pro or con.
5.   Occasion: Breakfast meeting at 7:30 A.M. Speaker (only one) at 8:00. Some sleepy, some still eating. Other weekly speakers primarily inform, but some entertainment is expected. Thirty-minute time limit for talk and questions is rigid.
6.   Implications for presentation:
   a.   Need strong attention-getter.
   b.   Keep it light, work in anecdotes.
   c.   Emphasize scientific aspects, practical applications.
   d.   Talk twenty minutes, be ready for questions.
   e.   Expect some interrruptions, disorganization.

Building the Informative Presentation

Individual informative presentations commonly fall into three basic categories: instructions, reports, and lectures. The functions of oral reports and instructions are much like those of written reports and instructions, so be certain to check Chapter 4 for help in preparing instructions and Chapters 9 and 10 for help with reports. Lectures include most other informative presentations you may give. You may be explaining how something works or trying to explain some new technical theory or concept, for example.

The primary difference between the content of a written informative presentation and an oral presentation on the same basic materials in the scope that can be covered. You simply cannot cover as much information orally as you can in writing. For every point made orally you need much more supporting detail to clarify, exemplify, reinforce, and just to allow time for the point to sink in. Three or four good, fairly complex points is about all you can expect to make in a typical fifteen- or twenty-minute presentation. In longer presentations the ratio of new points to minutes is even smaller. So remember to keep the number of new points small and to reinforce each one with plenty of details.

Since developmental details for various kinds of informative presentations are discussed throughout this text, we need not consider them here. You can use the suggestions developed for written communications quite effectively if you remember to add an extra example, a few more concrete details, another analogy.

Begin the actual preparation of your speech by developing the main body; you can add an introduction and conclusion later. Develop a rough outline much as you would for a written presentation. Use this outline to organize your material, and change it when necessary as you accumulate material. When you think you have enough material, you can prepare a detailed formal outline, arranging your points in a logical sequence. If you are going to read your presentation, go ahead and write out your script from the outline; if you are using the more common and more often recommended method of speaking from notes, you can use your outline as a basic structure for your notes.

With the body of your presentation carefully outlined, you are ready to develop the two sections that can make or break your presentation: the introduction and the conclusion.

Introduction ● The introduction of an oral presentation is considered by many experts to be the most critical section. Certainly it is important. If you do not establish rapport with your listeners in the first few seconds, you cannot expect them to fully absorb and appreciate what you say for the next few minutes. Even though the overwhelming purpose of your presentation is to inform your listeners, you must include holding their interest as a necessary secondary purpose. And you cannot hold their interest unless you first get it.

Your listeners will probably have some interest in your presentation to begin with. They will want to learn more about the subject because it is important to them or because it is timely and relevant. They may have a somewhat forced, but nonetheless real, interest because they know that they will have to use the information. But this initial interest can fade rapidly unless you make a strong beginning to solidify that interest and then work to maintain it throughout.

Most texts on public speaking offer lists of devices you can use to get an audience's interest. As you speak more and more you will no doubt want to expand your supply, but the five techniques listed below will give you a good base.

1.   *Referring to the subject, occasion, or audience:* When your subject is compelling and your audience is eagerly anticipating what you have to say, you are often wise merely to announce your intention and plunge ahead. This also shows the audience just how interested you are, and how eager you are to share your information with them. A long-awaited report or instructions for using an important new piece of equipment can often begin this way. For instance, you might begin a report on a successful fund-raising effort by saying: "We made it—and then some—120 percent of our goal! You'll all keep your jobs—except, of course, those of you who'll be promoted. Now, for the details."

Referring to the occasion or audience works much the same way. If the occasion is, indeed, something special, you can say so and then go right ahead, showing how your presentation fits the occasion. If, for example, you are explaining the operation of some new facilities, you might begin by saying, "We're all here for the same reason: to learn how to operate the half-million dollars' worth of equipment standing next to me. I'd like to say that it's simple to run, but it isn't. So let's begin a careful look at just how it should be operated."

2.   *Startling or shocking the audience:* You can often get the absolute attention of your audience if you begin with something totally unexpected. It may be controversial, contradictory, outrageous, or merely unexpected. The point is that the audience snaps to attention to learn what is going on. You can then proceed to tell them what is going on and to fit the remark into the context of your presentation. For instance, an after-dinner speech on contaminants in food might begin with the following remarks: "You have all just been poisoned. The roast beef, potatoes, pie, even the water, all contained poison. And not just one poison—arsenic, sure, but also mercury, DDT, parathion, and lead oxide."

3.   *Giving a vivid example:* One of your best introductory devices might otherwise be buried somewhere in the body of your presentation. If you have collected some effective examples to illustrate points in your presentation, you can often use one of those examples to open your speech. This is an especially useful technique if your example is itself shocking, humorous, or emotionally appealing. You might, for instance, begin a speech explaining hyperactivity in children by giving a brief, typical case study, emphasizing the problems the child's behavior caused him, his family, and his teacher, then showing how he was successfully treated.

4.   *Using a quotation:* Quotations from well-known people can be very effective in gaining an audience's interest, providing that the quotation itself is interesting, the original speaker's name is familiar, and you can show the quotation's relevance to the main point of your presentation. Assume, for example, that you are instructing some new employees in

effective customer relations. You might begin like this: "You're all familiar with Will Rogers's famous statement, 'I never met a man I didn't like.'
Don't you believe it. You'll meet plenty of men you don't like, and
women, and children."

5. *Relating an interesting anecdote:* Nearly everyone responds to humor. A
good joke or a clever story will get the attention of almost any audience.
But such an opening can be dangerous. If the story is not at least mildly
entertaining, it will do more harm than good; and if it is not made relevant to your presentation it will distract your listeners from your
important points. When you find a good story, use it. Look in joke books
and other sources too, but don't try too hard. Use some other beginning
rather than a dull or irrelevant story. Notice in the anecdote below, used
to begin a lecture on resumes, that additional audience appeal is gained
by making the speaker the butt of the story.

> We all have skeletons in our closets. None of us could truthfully present a per-
> fect resume, without one or two minor blemishes. I was absolutely paranoid
> about my own when I first started job hunting. I'd done quite well in school—
> except for three quarters immediately following my twenty-first birthday, when
> quite frankly I spent all my time pursuing the wrong kind of extracurricular
> activities and could have been kicked out of school if my previous grades hadn't
> been so good. Well, finally the day came. An interviewer sat across the table look-
> ing at a transcript of my grades. "Hm, very good grades. Hm . . . what? Hah!
> Hah! Hah! Was it fun?" He went on to explain that my earlier, and especially
> later good grades were enough to sell him. He'd allow me the one slip.

Remember, too, that the test of an anecdote is in the telling. Try it out on
someone before you use it in a speech. It may sound good on paper but not so
good live. Or it may not look so promising on paper but work very well when
properly told.

Getting the listener's attention is the primary function of a good introduction, but there are two other important functions: You must build a
transition into the body, and you must show the importance or relevance of
your information. Show the listeners why they need your information. As implied above, a strong opening is useful only if it can be related to the main body
of the presentation. Furthermore, since your presentation will be primarily informative rather than entertaining, you must get your listeners involved in the
main purpose or they will quickly lose interest. Here is how two of the attention-getting devices mentioned above—the quotation from Will Rogers and the
shocking statement about poisoned food—can be tied into the body of a speech:

> . . . There is no call for me to stand here and tell you that all of your customers
> will be cooperative, pleasant, and understanding. I'd be lying, and you would
> know it. The vast majority will be pleasant to work with, but the other kind do
> exist. Most of the battle will be a matter of exercising your self-control, of biting
> your tongue, of smiling when you feel more like screaming or crying. However,

those of us who have been through the wars for a while have come up with some procedures to make your job a bit easier, to help you to cope with the occasionally impolite or unruly customer. Let's consider a few.

... No, you will not die immediately; we've done nothing unusual to the food. The fact is, everything you eat contains these and many other poisons. People in the United States and throughout the world are systematically poisoning not just the air, the waterways, the earth, but the very food they eat. I'd like to present tonight some information about just how serious and widespread this problem is, then offer some suggestions on what can be done about it.

Conclusion ● The conclusion is your last chance at the audience, the last thing they will hear before they get on about their business; and most importantly, it is likely to be the part of the presentation that they remember the longest. So make sure that you close strongly. If you have made three or four distinct points, you can use a closing summary, restating and reemphasizing each point. If everything you have said has been in support of one central idea, restate that idea clearly and forcefully. You can still sum up if you wish, but concentrate on a forceful reiteration of your main point. Many speakers will even save a particularly vivid example or detail to use with the last statement.

If you are available immediately for questions, by all means say so. If not, you can suggest where, when, and how listeners can get more information. If you wish to show your appreciation for being invited to speak, you can do so, but do not bother with the often hollow "thank you" used in closing many presentations. Here are possible conclusions relating to the two introductions given above:

... So, remember those three points. Concentrate on what the customer is saying, not on the manner of expression. Consider the situation. Even if the problem is entirely of the customer's own making, it is a problem. Walk away or get a supervisor if the customer becomes obscene or threatening. But, most of all, remember that most customers are pleasant, polite, and frankly delightful. You'll have many more good times than bad.

If all else fails, one of your supervisors or I will always be handy.

... Once again, I'd like to assure you that we have not reached the point of no return. The situation is grave but not hopeless. So follow the suggestions I've passed on tonight. Devise your own solutions. As a well-educated group who solve technological problems daily, you are uniquely qualified to come up with new solutions. Persevere.

A Final Checklist

1. Will your presentation hold your audience's interest? No matter what your primary purpose, you cannot achieve it unless you can keep your audience interested.

2. Do you know your subject well enough? If your knowledge is skimpy, perceptive listeners will notice.

3.   Are your beginning and ending strong? They are the critical points.
4.   Is your planned presentation geared to the speaking situation as well as to the listeners? Time and place must both be considered.
5.   Has each of your points been developed with sufficient detail? Give each point time to soak in.
6.   Are your ideas organized in some logical pattern? Be sure also to build transition between points.
Following is a typical outline for a short informative presentation.

Bonsai: The Japanese Art of Growing Miniature Trees

Introduction

1.0   Reference to group and occasion
     1.1   An honor to speak before all of these future horticulturists
     1.2   Happiness particularly at your interest in Bonsai

Body
2.0   Learning about Bonsai a wise choice
     2.1   Simple, inexpensive to create and care for
     2.2   Well-suited for apartments, dormitories, etc.
3.0   Three main focuses of presentation
     3.1   Creating
     3.2   Training
     3.3   Caring for
4.0   Creating a Bonsai
     4.1   Preparing the container, tools, and soil
     4.2   Establishing an essential form
     4.3   Root pruning and potting
5.0   Training the established Bonsai
     5.1   Wiring the branches
     5.2   Pinching back
     5.3   Repotting as needed
6.0   Caring for the tree
     6.1   Watering
     6.2   Feeding
     6.3   Pest control

Conclusion
7.0   Only limits—your creativity and energy
     7.1   Unlimited variety of types
     7.2   Need for care reiterated

Delivery

The first section of this chapter told you that if you prepare your presentation wisely and thoroughly you can give it successfully even if you are not a skilled, experienced speaker. This is certainly true. More presentations are ruined through ineffective preparation than through ineffective delivery. But delivery is obviously important, and this section is designed to help you to improve your delivery, especially in your first appearances before an audience.

Choosing a Method of Delivery

1. *Impromptu:* To put it bluntly, this means "winging it," just getting up and talking. And frankly, only a fool speaks impromptu when it can be avoided. You will probably have to speak without any advance notice sometime, and you will just have to muddle through as best you can. The important point is never to speak impromptu when you do have advance notice. Even if you learn you are to speak only five minutes in advance, put that time to use. Size up your subject and your audience. Sketch out some rough notes. Look for an introductory device. Then deliver your presentation as calmly as you can. Do not hesitate to look at your notes or to pause to consider what to say next. Generally, the audience will realize your situation and will be tolerant of some awkward moments. Even if they are not, you will still do well to go slowly and to make sure that everything you need to say gets said and said clearly.

One final note to consider here is the relationship between stage fright and method of delivery. Surprisingly, many of the worst, least effective presentations are given in low-pressure situations by a speaker who is not the least bit nervous. This lack of nervousness is a problem because it lulls the speaker into failing to prepare. Result: confusion, mistakes, omissions, unsuccessful communication. Take all oral presentations seriously; prepare thoroughly; speak impromptu only when you have no alternative.

2. *Memorized:* This method is physically possible but most definitely not recommended. When you memorize and recite word for word, you risk almost total failure and can at best give a mediocre presentation. Very few people can recite an entire presentation from memory without sounding wooden or mechanical—sounding, in other words, as if they were reciting.

If you have memorized word for word, what happens if you forget a word? Do you pause trying to get your mind back in the groove? And if you miss the groove, do you back up and repeat part of a point or jump ahead to the next major point? These are real possibilities, and they can destroy your presentation. Do not memorize verbatim.

3. *Read:* The read presentation is just that. You write out a script and read it to your audience. While not as effective in most cases as the fourth method, extemporary delivery, it can be quite effective when done properly in the right situation. Consider reading your presentation in the following instances:

    a. Your content is so technical and complex that it would otherwise be difficult to keep straight. This also applies when your content is heavy with statistics or direct quotations.

Nearly all effective oral presentations are delivered extemporaneously, a method which enables you to speak expertly and naturally, with inflection in your voice and at a natural pace.

b.　You will be timed exactly, as in a television or radio taping.

c.　You are likely to be quoted or reported in the news and an inaccurate statement might be harmful. This is sometimes true with important public announcements or position statements on public issues.

d.　You are asked to submit a written copy of your presentation for distribution. Reading your material is the only way to make certain that the oral agrees with the written text.

　　When you do decide to read a presentation, prepare carefully and practice. With some hard preliminary work you can make your actual presentation almost as effective as if you had presented it extempore. In fact, the best comment you could hope for about a read speech is "It didn't sound as if you were reading." Here are some suggestions for effective read presentations:

a.　Prepare a script word for word, and stick to it. Do not ad lib.

b.　Check the language of your script. Make it sound like spoken English, not written English.

c.　Make a copy that is easy to read from. Type it triple-spaced or print it by hand in large letters with large spacings. Underline and add marginal comments if that will help.

d.　Practice, practice, and practice some more. Get to where you could almost recite the speech. Then practice reading with inflection in your voice, pausing for effect just as you would if speaking extemporaneously.

e. When you actually deliver the speech, read it, but try to maintain at least some eye contact with your audience. Look at your script to pick up a sentence or clause, then say it to your audience.

4. *Extemporaneous:* Nearly all effective oral presentations are delivered extemporaneously. Contrary to what many people believe, this does not mean just getting up and talking. Rather, it means preparing your material carefully, practicing the presentation until you know your material well, then delivering the presentation using notes. Such a presentation enables you to speak effectively and naturally with inflection in your voice and at a natural pace. Yet it assures that you will cover your material fully, accurately, and in the proper sequence. It also minimizes your chances of forgetting what to say next. In short, this method has the advantages of all three other methods, without most of their liabilities. Unless you absolutely must speak impromptu or you are certain that reading is necessary, speak extemporaneously. The remainder of this section is designed to help you to deliver your extemporaneous presentation more effectively.

Using Your Voice Effectively

One of the most obvious characteristics of an effective speaking voice is its pleasing tone—it sounds good. If you naturally have a pleasant-sounding voice, you have a built-in advantage. If your voice often sounds harsh, nasal, squeaky, husky, or breathy, you may never perfect the mellow tones of professional radio or television announcers, but you can improve. Speech texts can suggest a number of exercises to help you to eliminate specific problems. However, practice in speaking clearly and distinctly, especially using a tape recorder, can help almost all of us.

A second characteristic of a good speaking voice is intelligibility. If your listeners cannot understand you, you obviously are not communicating effectively. Check first to adjust the volume of your delivery. You are, no doubt, already aware if your voice is unusually loud, excessively soft, or about average. If it is louder or softer than most people's, you will have to practice at toning it down or at projecting more forcefully. Next you will have to take into account the situation of the proposed speech so that you can adjust your volume level accordingly. Be sure to consider any expected outside noises such as busboys cleaning tables or listeners clinking glasses.

In addition to your volume, check your pace and pronunciation. Using a tape recorder or a friend for feedback, practice speaking clearly and carefully, pronouncing each syllable distinctly. Often this means slowing down, but merely speaking slowly will not guarantee clarity. Make certain too that you can correctly pronounce any unfamiliar words you may be using.

Even a clear, pleasant voice can be ineffective if it constantly speaks in a dull monotone. Practice varying the pace of your presentation. The average pace is somewhere around 150 words per minute, but you should sometimes talk much more slowly and sometimes more rapidly. Also, you should deliver some points extra softly and some extra loudly. Your voice should be sometimes calm, sometimes forceful, sometimes low pitched, sometimes higher pitched. Practice till you sound like someone talking, expressing emotion in your voice to fit the emotion in your content. Avoid excessive, prolonged emotion; try to sound natural.

Using Your Body Effectively

Inexperienced speakers too often overlook the importance of their bodies, concentrating solely on developing their voices. That is unfortunate, because an audience will react as much to what it sees as to what it hears, even if what the speaker is saying is absolutely spellbinding. Therefore a facial expression, a wave of the hand, or even a shrug of the shoulder will communicate as much as many words.

Studies indicate that listeners react more strongly to a speaker's apparent perception of them than to any other aspect of presentation. If you appear to dislike them or to be unconcerned about them, they will dislike you and will react negatively to what you say. If they can sense, though, that you are genuinely interested in them and are trying to communicate with them, they will automatically give you any possible benefit of doubt. Nothing alienates an audience more than a speaker who is apparently oblivious to their presence.

Eye Contact ● The best single thing you can do to give your audience the proper impression is to look at them, to talk with them as individuals. You obviously cannot look into the eyes of every member of the group constantly, but you can single out each member occasionally. Look at one person for a few seconds, then move your eyes a bit and talk to someone else. During the course of your presentation you can talk with most members of a small audience; with a larger audience you can single out members in every corner of the room. Be especially careful to avoid staring too long at one person or area; avoid also the opposite problem of merely moving your head around. Actually look at everyone.

Body Movement ● Find a comfortable, unobtrusive posture. You should neither stand in a rigid brace like a West Point plebe being inspected nor sprawl over the podium or speaker's table. Move around a bit. Change positions. Once you learn to relax while speaking, you can be natural. Even the first few times out, though, you can move a bit. Come around to the front of the podium to make an important point. Bend forward a little to speak confidentially.

Striking a comfortable balance is the important concern. Move around, but avoid distracting tics or eccentricities. Do not aimlessly tap a foot, sway from side to side, tap your fingers on the podium, or shuffle your notes.

Gestures ● You can gesture with your hands, your face, even your shoulders. If you smile when saying something happy, frown at something distasteful, and even leer and scowl to match the words you are using, you will greatly enhance the effectiveness of those words. Do not try to practice facial expressions. Just let them come naturally. Smile, scowl, or whatever when you feel like it, and the audience will feel that way too.

Hand and arm gestures fit into two basic varieties: descriptive and conventional. Descriptive gestures are those that seemingly come naturally: holding your finger and thumb apart to show "It was this close," or demonstrating with hand movements a method of connecting two pieces of apparatus. Conventional gestures are those that do not actually describe something but that have a certain meaning because they are widely used to convey that meaning. Common examples are the clenched fist to show anger or determination, the pointed finger to challenge or accuse, or crossed arms to show disgust.

If you expect to be speaking frequently, you can practice conventional and common descriptive gestures, but be careful about practicing them to use in particular spots in a given speech. Artificial gestures will detract from your effectiveness. Your best course will be to learn to use gestures, then use them spontaneously in your presentations. Be especially careful to avoid the two deadly extremes: using no gestures at all and flapping your arms wildly about. Moving your arms consistently will distract the listeners' attention from your words and may also make you appear foolish. Standing with your arms locked behind you or dangling lifelessly at your sides or even tightly gripping the podium will rob your presentation of vigor and help to put your audience to sleep.

Using Words Effectively

It is when you come to the critical task of selecting exact words to use to express your key points that the differences between writing and speaking become most important. While many of your concerns will be the same whether your presentation is to be spoken or written, several differences between written and spoken language are significant.

Reading or listening, almost everyone receives many more messages (sometimes called *noise*) than he or she is specifically listening or reading for. The brain can process many more messages from the senses than a speaker or writer can provide. However, this phenomenon is even more important to the speaker than to the writer. To begin with, we do not speak as rapidly as the average person reads. If we tried, no one could understand our words. So the amount of noise heard by the listener is almost certain to be higher than that perceived by the reader. Second, the reader who misunderstands or is distracted can always reread. The listener cannot usually re-listen. The implication should be clear. You must do everything possible to keep the listeners' attention, to help them shut out the noise.

The second major difference has already been implied: You must be clear, thorough, and understandable the first time around. Otherwise, your listener cannot understand you and follow your reasoning.

The final difference of significance here is the lack of punctuation and special printing techniques in the oral presentation. You cannot use periods, paragraph indentations, underlines, or headings to indicate your organization or to give emphasis. Instead, you must rely on your voice inflection and, most importantly, on your choice of words.

Precision ● Since your listener can neither refer back to a previous sentence nor pause to consider a word or expression, you must be absolutely clear and precise. Avoid vagueness and ambiguity, of course, but also make doubly certain that you give sharp, concrete, precise images. Use "IBM 370 series" not "second-generation computer"; use " ±0.001" not "standard tolerances."

Imagery ● To help increase your listener's attention and comprehension, build concrete word pictures. Give descriptive details and use words with clear, vivid connotation. Rather than saying "a pair of old sneakers," you might say something like this:

> . . . a pair of once-white, high-topped Converse All Stars that were at least ten years old. They were held together with black electrical tape, which had been there for some time and allowed the left little toe to peek through. The laces, knotted in at least three places each, were a vivid orange.

While you certainly cannot give this much detail throughout your presentation, you can select key images that you want your audience to focus on and develop good word pictures for each. And throughout you can use words with sharp, effective meaning.

Technicality ● It should go without saying that speaking at the appropriate level of technicality is essential. More even than your written communication, your oral communication must be aimed at the appropriate level of technicality. If you use words your listeners do not understand or refer to equipment or materials they have not heard of, you will not communicate effectively. Imagine, unless you are a respiratory therapist or have a great deal of experience around respirators, what meaning you would get from *MA-1;* or, switching fields, try *variable annuities* or *heat sink.*

When you have the slightest doubt whether an audience will understand a term, avoid it. When you know of two ways to express an idea, pick the simpler, less technical way. You need not go to the ridiculous extreme of talking down to your audience or of using an elementary-school vocabulary; use your own good judgment.

Appropriateness ● Inexperienced speakers often have trouble finding a tone which is neither too formal nor too casual for the situation. Make your presentation sound like spoken, not written, English, even if you are reading it. Use your normal speech patterns, being careful only to be clear and thorough. Avoid slang, jargon, and offensive terms, both for appropriateness and clarity.

If the speaking situation is a bit more dignified and formal, be an extra bit careful. If the situation is extremely casual and the audience familiar, adjust accordingly. Use good, clear, spoken English without any frills or extremes, and appropriateness should be no major problem.

Transition ● Since you cannot use paragraphs, headings, or other typographical devices to indicate your organization, you must signal with words each new point or subpoint you begin. One good device to help you give these signals is a summary. Many speakers use a preliminary summary immediately following their introductions to indicate the main points they will be making. A speaker addressing a group of Midwestern farm managers on new developments in meteorology might summarize the main points this way:

> There are three recent advances you might be interested in. First, I want to tell you about our new, easier access to satellite pictures. Then you might want to know more about the improved long-range forecasting techniques that have been reported by the media recently. Finally, we can consider the military's recent advances in cloud seeding and storm diversion. It's still experimental, of course, but it looks promising for the not-too-distant future.

Summaries also work effectively at the conclusion of a presentation. You can simply restate two or three key points to help the readers remember them.

But merely summarizing key points does not automatically signal where one point leaves off and another begins. To do this you must use transitional expressions. Each time you complete a point, use an appropriate expression to indicate that you are ready to move on to the next one. Here are typical transitional expressions:

> . . . Satellite photographs can never really help us forecast more than four or five days ahead. However, we have made several recent advances in longer-range forecasting. One especially promising technique is. . . .

> . . . Now that you've heard how we predict the weather, you'd no doubt like to hear what we are doing about it. No, we cannot control the weather yet, but. . . .

Using Visual Aids

Just as illustrations are integral parts of many effective written presentations, so are visual aids integral parts of many oral presentations; they serve the same basic functions. A well-chosen slide, transparency, picture, poster, or any other appropriate object can be used to clarify a point, to make your presentation more concise, or to make a point more vivid. Many of the same charts and graphs you would use in a written presentation can be made into slides, transparencies, or posters. Drawings and tables can be similarly reproduced. But you can also use many objects that you could never include in a written presentation. If you are discussing small tools or pieces of equipment you can ac-

A well-chosen slide, transparency, picture, or poster can be used to clarify a point, to make a presentation more concise, or to make a point more vivid.

tually bring them before the group. Scale models or samples can also be used effectively.

Do not use visual aids just when you find them necessary. Look for chances to use them. The visual aids will help involve your listeners' sense of sight. By giving them something concrete to look at, you minimize the chance of their eyes and minds wandering. You keep the audience's interest focused where you want it focused.

The following suggestions will help you to use visual aids more effectively. Remember, though, with visual aids as with any mode of communication, do what your judgment tells you will work best in the particular situation:

1. Be sure your material is large enough to be seen and read clearly from all parts of the room.

2. Stand out of the way when handling or pointing to visual aids, so that your body does not block someone's view.

3. Make each item as simple and clear as possible. Do not try to impress your audience with unnecessarily detailed aids.

4. Present equations and similar data only when necessary, and then present it in the simplest form you can. Do not expect your audience to sit quietly studying your data for long periods at a time.

5.   Show your visual aids only when you want your audience to look at them, and put them away as soon as you move on to another part of the presentation. If the audience sees strange devices or even posters near the speaker it cannot help but be distracted.

6.   As part of your routine preliminary planning, check the speech site for adaptability to your intended visuals. Make certain of the necessary equipment and cords; check on room darkening and electrical outlets. Make sure that you will be able to use the visuals you want in the manner you want.

Some Special Situations

Thus far, the primary emphasis of this chapter has been on planning, preparing, and delivering an informative oral presentation. The vast majority of job-related oral presentations are informative rather than persuasive or entertaining, but many technical people do occasionally have to give other types of presentations. This section will offer suggestions about two such presentations—the persuasive speech and the goodwill speech.

The Persuasive Speech

Just as typical technical professionals occasionally have to write something designed primarily to persuade rather than inform, so must they occasionally deliver an oral presentation whose basic purpose is to convince the audience of something. Some fields, such as sales, obviously require more persuasive presentations than others, but you can be sure that if your position requires a significant amount of public speaking at least a few of your presentations will need to be persuasive. You might, for instance, have to persuade your company's executives to fund the changeover to a new technique; then you might have to encourage your subordinates to give the technique a fair chance to succeed. Later you might have to defend the technique to community groups or even governmental agencies. Finally, you might want to convince your colleagues in a professional organization that they should adopt the technique themselves.

Persuasive speeches come in as many forms as do informative presentations, with each audience, each situation making its own demands. However, many persuasive presentations have certain aspects in common.

Many effective public speakers have found that a particular five-step pattern works well in nearly all persuasive speeches. Sometimes called the motivation sequence, this pattern can be adapted to almost any situation calling for a persuasive speech.

Attention Step ● First, as in any oral presentation, you must get your listeners' attention. Attention-getting devices were discussed earlier in the chapter.

Need Step ● Next, show your listeners that they need the information you will present. Show them that a problem exists, that the current situation needs improvement. You might, for instance, point out that your company's sales are down 19 percent from last year's, or that your work load has increased by 23 percent recently with only a 5 percent increase in staff, or that in spite of numerous technological advances in recent years you are still forced to use the same reporting methods you used when the department was organized thirteen years ago. Convince your listeners that a definite problem exists, and the rest of your task will be much easier.

The real key to developing an effective need step is audience analysis. Study your potential audience carefully, looking for common needs that you can appeal to. Look first for needs unique to them as a group—as firefighters, Rotarians, or electrical engineers. Then look for needs they share with others, basic human needs like food, security, freedom from chaos, affection, acceptance, self-esteem, and status. Find at least one and preferably several needs to appeal to. The more important the need is to your audience, the more effective your presentation.

Make your presentation of the need or problem thorough and vivid. Begin with a clear, concise statement of the problem. Give at least one good example to clarify and vivify it. Then use whatever factual evidence you need to support your contention. Emphasize throughout that the problem directly affects the listeners. Make the need real and concrete to each one.

Satisfaction Step ● The next step is to present a solution to the need or problem you have just created. This is, of course, the heart of your presentation. What you present here as the best or possibly even only solution to a serious problem is the central idea behind your entire speech. The need, in fact, was established specifically to be filled by this step.

The basis of this step is a clear, concise statement of exactly what action or attitude you want the audience to adopt in order to solve their problem. After stating your solution, be sure to explain it carefully and fully, using examples, statistics, and any other tools you need. It is often effective to retrace the key aspects of the problem presented in the need step and to show specifically how your proposed solution would satisfy each aspect. Consider also any major objections to your proposal. Determine what aspects of it may be unpopular. Then you can give evidence and examples to negate the objections before they actually arise.

Visualization Step ● Next you need to intensify the audience's desire for your solution. Show them the positive effects of accepting your proposal. Show them how much better things will be once your proposal is put into action. The need step pointed out an unpleasant current situation; the satisfaction step showed how to improve the situation; the visualization step will show a pleasant situation resulting from the satisfaction step.

This step may include negative projections as well as positive. Show the reader how much worse the already bad situation might become if your suggestions are not accepted. Or contrast two future projections, one a positive look at the situation if your ideas are accepted, the other a negative look at the situation if your ideas are rejected.

The visualization must be both vivid and believable; you must make the future projection concrete and real to your audience. Use specific details and examples, and look for good visual aids. Suppose, for instance, you are trying to persuade your local county commissioners to donate five acres of unused county land for use as a children's playground. After getting the commissioners' attention, you show them two distinct needs: the need of area children for a place to play other than the streets, and the need of the citizens of the county to have their county-owned land put to some worthwhile use. Then you show them in your satisfaction step how donating the land to your nonprofit parents' association would solve both problems. Then, in your visualization step you can describe in detail what equipment and facilities you will have in the playground. You can give specific projections of the number of children using the playground at various times. Best of all, you can use visual aids. Slides, sketches, and photographs would all be helpful. You might show what the projected playground would look like and possibly what a similar playground somewhere else looks like, complete with happy children. Then you could show pictures of the vacant area now, along with pictures of area children playing in the streets, remembering to remind the group that by not accepting your proposal they would be permitting the land to remain idle and the children to continue playing in the streets.

Action Step ● The action step is your chance to turn the intellectual commitment you gained through the first four steps into definite action. Merely getting your audience to agree with you will seldom do much good unless you can get them to act on that agreement. With each hour that passes, listeners will lose interest in your proposal, so the time for action is now. Tell them exactly what they can and should do. And explain how they can do it. Then shut your mouth. Do not belabor this step. Boring the audience now could seriously lessen the impact of your entire presentation.

In the hypothetical presentation about the playground just mentioned, your action step might begin with the explicit statement that you would like the group to vote as soon as possible, preferably at that meeting, to transfer the land. Then you could explain that your group could begin constructing the playground within thirty days and could have it ready for use within another thirty. You could close your entire presentation by saying something like this:

> Within sixty days from tonight we can have our younger children off the streets, our public land serving the public. We ask for your immediate action.

Some guidelines:

1. In order to change someone's mind, you must get his or her undivided attention. Develop the best introductory device you can.

2. No matter how attractive your solution, the audience will be hesitant to accept it unless they are first convinced that the problem you point out is both serious and immediate. Make the problem relate specifically to your audience.

3. Double-check your satisfaction step to make certain that it clearly and fully presents the best possible solution you can find.

4. Use concrete examples and visual aids throughout if you can, but try especially to use them in your visualization step.

5. Make your call to action clear and crisp, and do not forget to show your audience how and where to take your suggested action.

6. Double-check your logic as closely as you would for a written presentation. Glaring fallacies in your reasoning will destroy your presentation.

7. Take special care in preparing your presentation if you are going to be facing a hostile audience. If the audience disagrees with you, you will need many more supporting facts and examples than you would for a neutral audience. Also, be careful to avoid insulting or otherwise offending those who differ with you.

8. Be open and direct. Subtle "snow jobs" are sometimes effective, but they require great skill to handle successfully. Even should you subtly cajole your listeners into siding with you, they can easily be persuaded to change their minds again. Worst of all, if they learn they have been fooled or manipulated, they will never again trust you.

Below is a brief outline for a typical persuasive presentation. Study it closely, but remember that each speaking situation is unique.

A Unified County Fire District

Attention Step	I.	Description of Anderson fire, October 1981
		A. Six deaths
		B. Fire station three blocks away
Need Step	II.	Last year's thirty-five unnecessary deaths
		A. Areas in no fire district
		B. Areas miles from their station with other stations close by
		C. Many citizens who do not know what station to call
	III.	Last year's wasted money
		A. Duplication of equipment
		B. Low-volume prices
		C. Unbalanced personnel distribution
Satisfaction Step	IV.	Combining thirty-five districts as a solution
		A. Everyone in county covered

B. Appointed chief
C. Present firefighters all retained
D. Stations and equipment retained
V. Experiences elsewhere
    A. Nationwide statistics
    B. Seminole, Florida, as a typical example
VI. Minimal drawbacks
    A. No immediate tax increases
    B. No one loses job
    C. No loss of protection in wealthy areas

Visualization
Step

VII. Worsening problems if change is not made
    A. More needless death and destruction
    B. Distortion of taxes
VIII. Brighter future with change
    A. Parents sleeping securely
    B. Taxpayers getting their money's worth

Action Step   IX. Vote yes on November 7

The Goodwill Speech

As discussed in the opening section of this chapter, all oral presentations have persuasive, informative, and entertainment aspects even though they are generally identifiable as one of the three. Nowhere is this multiple-element nature more apparent than in what is commonly termed the *goodwill speech*. If we had to put the goodwill speech into one category, we would probably call it informative, but persuasion and entertainment are essential to any effective goodwill presentation.

Goodwill speeches are those presentations delivered by technical or professional people to outside groups. Civic groups, professional groups, and school groups, for example, commonly ask for this type of presentation.

The apparent goal of a presentation to such a group is informative: to give the group useful, relevant information about some aspect of your work. But in addition to this, there is the equally important goal of persuasion, of impressing the group with yourself, your company or organization, or even your profession. Entertainment is important, not as a distinct goal, but as a means of helping to achieve other goals. This is especially true with presentations given at luncheons, banquets, or special assemblies. The listeners expect information, but they also expect to enjoy themselves.

Plan a goodwill speech essentially as you would any informative speech. Use the basic informative structure shown before, or vary it to suit your needs. Select a subject carefully, looking for one that is new or unique. Try to find aspects of your work or your organization that are not generally known. Analyze your audience carefully and make your subject matter as important to each listener as you can. What you see as important about your job does not always seem that way to someone else.

Develop the level of technicality of your material carefully. Give specific details and examples, emphasizing if you can any inside information you can offer. But do not speak over the heads of your listeners or let them get mired down in a mass of detail.

Show the audience specifically how you can help them or how you are already helping them every day. Show them, too, how you can be of further service. However, do not under any circumstances seem to be merely bragging, telling them how wonderful or indispensable you are. You are trying throughout to show your listeners that you, your organization, and your profession are, indeed, marvelous. But let them decide that on their own.

Select introductory devices, illustrative examples, and visual aids that are vivid and interesting. Humorous anecdotes are especially effective. Just remember, though, that you are not expected to be a stand-up comedian.

Two final notes should be considered: prepare carefully to cope with clinking dishes or other such distractions if your speech is to be given after a meal. Be prepared to answer questions, not only concerning your speech content, but also about other aspects of your job or organization.

Following is an outline for a typical goodwill speech. This particular speech was given to a typical civic club by a discount store personnel manager.

Hiring Clerks for the Big M

Introduction

1.0　Anecdote about mixing mange cream and Tough-Skin
　　1.1　Some people who are hard to get along with
　　1.2　Learning to smile and "turn the other cheek"
2.0　Discount houses bringing out the worst in people
　　2.1　Effects of low overhead
　　2.2　Effects of self-service
　　2.3　Effects of tight security

Body

3.0　What I look for
　　3.1　Must handle many customers
　　3.2　Must work unusual hours
　　3.3　Must smile and be courteous
　　3.4　Must show initiative and resourcefulness
　　　　3.4.1　Lost grandmother incident
　　　　3.4.2　Misfired-pistol story
4.0　How I find it
　　4.1　Paying top salaries
　　　　4.1.1　Salary statistics
　　　　4.1.2　Anecdote about high-school teacher
　　4.2　Gambling on former losers
　　　　4.2.1　Minor police records
　　　　4.2.2　Poor work records
　　　　4.2.3　Physically handicapped
　　　　4.2.4　Very young

Conclusion

5.0 The rewards
    5.1 Lucerne Street store manager
    5.2 Civic awards
    5.3 Brisk business
6.0 The best job I could ask for
    6.1 Never dull
    6.2 Full of satisfactions

• EXERCISES

1. Plan a ten to twelve minute informative speech for your classmates in this course. Do each step presented in this section.
    a. Select a suitable subject.
    b. Develop a general and a specific purpose.
    c. Analyze your audience.
    d. Outline the body of the talk.
    e. Build an effective introduction.
    f. Develop a conclusion.

2. Show what changes you would need to make in preparing a speech on the same topic to a group of business people and a group of junior-high-school students.

3. At a time specified by your instructor, present to the class the speech prepared for Exercise 1. Use at least one visual aid.

4. Select one of the proposals you wrote in Chapter 11. Analyze your audience and prepare a ten-minute oral presentation of the proposal.

5. Plan a fifteen-minute goodwill speech for a group of juniors and seniors at your old high school, telling them of life as a student at your college.

6. Adapt your speech for the high-school group to fit a group of community leaders.

13

Daily Informal Communication

Overview   You can become an efficient telephone user by following a few common-sense rules. Many kinds of unofficial communication provide a rapid, though often inaccurate, communication network within organizations. Become aware of what they can teach you.

Besides speaking and writing, we communicate nonverbally through posture, gesture, and facial expression. The wrong nonverbal signals may counteract what you are trying to communicate with words.

Finally, to round out your communication skills, give some attention to yourself as a listener. Train yourself to shut out both physical and mental distractions and to identify and remember important ideas.

The preceding chapters have been concerned with relatively formal communication. Writing letters, memorandums, or reports or giving speeches—all are similar in that the communicator is conscious of what he or she is doing, recognizes the need to communicate effectively, and can prepare accordingly. Such situations are often only a small part of daily communication. It is difficult to project just how much of the typical technical person's day will be spent in what type of communication, but it is likely that you will spend at least as much time on informal communications such as talking on the telephone or chatting with colleagues and supervisors as you will on the more formal types. And those phone calls and casual conversations are often every bit as important in conducting your organization's business and advancing your own cause as are the more formal situations.

Because of their almost unlimited diversity, the very informal situations cannot be examined as systematically as can more formal ones. This chapter, though, will offer some useful suggestions for improving your own informal communication skills.

Using the Telephone

Despite the development of on-line computers, telexes, and other mechanical means of rapidly communicating across long distances, the telephone remains our most widely used, most versatile tool for instantaneous long-distance communication. As a technical or business professional, you can greatly increase your job efficiency by improving your telephone skills. First you must learn when to use and when not to use the telephone. Then you must sharpen your skills both as an originator and as a receiver of calls.

Ineffective use of the telephone can lose customers, cost people jobs, and cause us all a great deal of work. Consider the three typical situations below:

Roy Ragsdale, chief X-ray technician at a large metropolitan hospital, decides to call the local distributor for much of the equipment used by his department. A new machine has been malfunctioning for several weeks and repairmen have been unable to get it working satisfactorily. When he reaches the switchboard at Bucksewco Products, he is asked to hold. After three minutes—it seems like fifteen to Roy—he is asked whom he is calling. He asks to speak to Olin Fischer, the sales representative he has been dealing with for years. The response from Fischer's office is "please hold." After a two-minute wait, a secretary informs Roy that Fischer is out of town, suggests he contact one of the other sales representatives, politely excuses herself, and hangs up. Roy subsequently begins doing business with Bales and Company distributors.

Christy King, assistant chief for operations of a city fire department, decides to call Mac Niblick, chief of the department in an adjoining suburb, to discuss coordination of coverage in a large shopping center located in part in each municipality. Mac answers the phone cordially, but does not bother to turn off the radio blaring a few feet away. While conversing he also stops to answer questions from several firefighters who step into his office. He also hollers at another firefighter to order his lunch. After many false starts, interruptions, and repetitions, Ms. King gets frustrated and gives up. Several months pass before the departments can get together again to discuss the matter.

Charlie Sanders is manager of Twin Oaks Acres, a large grain farm in eastern South Dakota. The farm's owner, Allison Salsberry, spends most of her time in Tucson, Arizona. Charlie makes all decisions in the day-to-day operation of the farm, but is expected to consult Ms. Salsberry before making any major decisions. Charlie is offered an excellent price for ten thousand bushels of wheat he has in storage. Charlie decides to call Ms. Salsberry about the wheat and several other matters. Ms. Salsberry is anxious to learn about other matters, not having been up to the farm in almost a year. She and Charlie chat for ten minutes. When Ms. Salsberry realizes how long they have talked, she asks Charlie if there is anything else he wants to discuss. Charlie forgets the wheat and hangs up. He sells the wheat without Ms. Salsberry's consent and later loses his job.

All of these situations involve more basic problems than ineffective telephone usage. They do, however, clearly demonstrate some of the ineffective telephone techniques commonly used by people who are otherwise effective communicators.

Determining When to Telephone

Many technical people make too little use of the telephone, writing or appearing in person instead. The phone can often be an effective substitute for either the written document or the face-to-face conversation. It offers advantages over either—and some disadvantages.

The telephone's most obvious advantage is speed; no generally available form of written communication can compete with it. Nor can the direct conversation. The phone's most commonly considered drawback is its cost, especially when long-distance tolls are involved. Common sense seems to verify this. A letter—a sheet of paper and a first-class postage stamp—certainly do not cost as much as the minimum rate for even a brief long-distance call to a nearby city.

The phone can often be an effective substitute for either the written document or the face-to-face conversation.

But what about time? Composing, typing, and posting letters requires time, and someone has to be paid for the time involved. Recent studies by various organizations determine the actual cost to an organization of preparing and sending an original letter at over five dollars. Your actual costs will vary somewhat, but the implication is obvious: a three- to five-minute long-distance phone call is generally less costly than an original letter.

A phone call also offers greater flexibility than does a written communication. The caller and the receiver can respond and react immediately; alternatives can be weighed, and unexpected difficulties can be identified and resolved. At the same time, the phone call is giving a personal touch that no written communication can match.

On the negative side, phone calls do not furnish permanent records or verification as do written documents. The need to "get it in writing" often dictates sending a letter.

Sending a written document is also recommended when the other party needs time to prepare a response. Even with the needed files or other materials at hand, he or she often cannot formulate an appropriate response while you wait on the other end of the line.

Written communication is also desirable when you wish to transmit the same information to a number of people. Form letters and duplicate memorandums cut the cost of written communication far below that of individual phone calls.

Also sometimes preferable to a phone call is the face-to-face meeting. Calls are more personal than letters or memorandums, but person-to-person meetings are even more so. When you want to make the strongest possible personal impression, do it by appearing in person rather than by calling.

Each mode—personal contact, phone call, written document—has advantages. Use them all. Make choosing among the three a natural part of your preliminary planning.

Improving Your Phone Techniques

The strengths of using the telephone are valid only when the telephone is used effectively. Unfortunately, many professionals whose other communication skills are keen have poor telephone skills. Using the telephone seems easy—after all, we have been talking on the phone since childhood—but using it well is not quite so simple. The suggestions given below and some practice should help considerably.

Suggestions for originating a call:
1. Plan your call carefully. Jot down the questions you want to ask or the points you wish to bring out. Gather supporting evidence and have it at hand.
2. State your business. Telling your listener early that you have three questions or that you wish to discuss four matters will save you time and will help to make certain that you get everything covered.
3. Identify yourself. Do not make your listener or his receptionist guess or ask who is calling.
4. Use good judgment about small talk. You should decide how much friendly chatting is desirable. Consider your relationship with the listener, the impression you wish to make, and the cost of the call.

Suggestions for receiving a call:
1. Take your cues from the caller. Let him decide how much small talk to make or what topics to cover. This does not imply that you cannot ask questions or bring up new topics. Nor does it suggest that you must listen interminably to a bore. Use good judgment and remember to give the initiator precedence.
2. Identify yourself and your title or department. If you cannot help the caller, switch him or her to someone who can.
3. Help the caller accomplish his or her goal. Answer questions fully and furnish any additional information you think will be helpful.
4. Give the caller your undivided attention. Excuse yourself from your companions to answer the phone and avoid conversing with them until the call is completed.

5. Keep interruptions to an absolute minimum. If you must leave the phone, explain clearly and fully why the interruption is necessary, and keep it brief.

6. Never put a caller on hold if you can possibly avoid it. Keep any necessary holds as brief as possible, and be certain to apologize to the caller.

Suggestions for callers and receivers:

1. Cultivate your telephone voice. On the telephone, your voice is you— no facial expressions or gestures to help. A cheerful voice conveys a smile; a bland, lifeless voice conveys disinterest or disapproval.

2. Practice courtesy whether you are calling, answering, or transferring.

3. Speak distinctly, with a normal conversational volume. Keep the mouthpiece approximately two finger widths away from your mouth.

Unofficial Communication

Throughout this text we have been examining on-the-job communication: formal, informal, written, and oral. The communication situations studied thus far comprise the majority of all official communication the typical technical person will face. However, they represent only a small portion of the actual job-related communication he or she will participate in. Unofficial communication in most organizations far exceeds official communication. The Friday afternoon bull session, the ten o'clock coffee break, the Saturday morning golfing foursome are significant vehicles for communication within an organization. No matter how effective an organization's formal, official communication structure, a complex informal and unofficial communication network is certain to exist.

Often criticized by managers as "rumor mills," grapevines are very real, significant aspects of all large organizations. Below are several examples of the grapevine in action.

Mary Crosby is a horticulturalist who manages the Greenville branch of Tarheel Nurseries, Incorporated. Upon returning from a meeting in Winston-Salem with company top management, she mentions to a tennis partner, Nancy Calcer, a local attorney, that a merger of Tarheel with Cape Fear Nurseries had been considered but rejected. Four days later she is asked by her secretary if it is true that the nursery will be closing next month. The following day a delegation of employees demands to see Ms. Crosby to verify the rumor that someone has bought out Tarheel Nurseries and will be bringing in all new employees.

Mac Sparrow, chief of a large city's police department, notifies his deputy chief, Dave Abernathy, to discontinue using one of the forms previously required several times per shift of all members of the uniformed patrol division. Abernathy passes word to Ron McLean, major in charge of uniformed patrol. McLean notifies his three shift commanders, who pass the word to division sergeants, who then notify the patrolmen. When, three days after the chief's decision, the patrolmen are officially notified by their sergeants, everyone who has not been out sick or on vacation already knows about the change.

Contrary to what many people believe, the grapevine is not necessarily detrimental to the organization. In fact, it often spreads information more effectively than formal channels do, especially in those "tall" organizations that have many layers of management or administration. In such organizations, word must be passed through so many levels that informal communications can often spread that word much more quickly.

Even "flat" organizations, having relatively few levels of authority, often benefit from informal communication systems. Flat organizations can generally communicate information vertically very efficiently; but the horizontal flow of information from one department to another is often slow. Informal communication systems can cut across normal departmental lines, thus speeding information flow.

As a manager, as a supervisor, or as a worker, you should always be aware of informal communication systems and use them wisely. The following observations are designed to help you to make better use of the grapevine:

1. It is ever-present but ever-changing. You cannot suppress informal communication without seriously damaging morale. Nor can you consciously manipulate it, because informal relationships are ever-changing, and the resultant communication flow is nearly impossible to predict.

2. Informal communication is notoriously inaccurate. The farther from the original source, the less accurate. Stories have been circulating for years about how communication gets distorted as one person hears a story then passes it on to the next. Such tall tales may never have actually occurred, but their message is accurate. Do not trust what you hear from the grapevine. Verify potentially important rumors. Feed accurate information into the system whenever you hear something you know to be incorrect.

3. Listen to casual conversation. In addition to actually gaining information from what is said, you can get a feel for morale. You can often identify possible conflicts and other serious problems before they break out.

4. Use informal communication to get at people's candid feelings and opinions. Many of us are more likely to speak freely and openly over a cup of coffee or a beer than in a formal interview or conference. This does not mean that you can consciously pump someone merely by buying two cups of coffee; it simply recognizes that we talk more freely when we are relaxed, and many of us do not relax well in formal, official settings.

5. Use the grapevine to facilitate the flow of information upward within the organization. Formal communication systems are designed primarily to transmit information down the organizational structure, with the return flow much slower and less reliable. Lower-level employees can

often communicate ideas and feelings informally that they could never get across through official channels.

6. Use informal channels to bypass possible roadblocks. Many organizations have one or two managers who simply do not communicate effectively, or who even block the free flow of information. The grapevine can easily get around such roadblocks.

7. Feed the grapevine with useful, accurate information. Learn who the key informal communicators are (seldom are they formally influential) and let them help you keep the flow of information going smoothly.

8. If you are a supervisor, keep your door open. Encourage direct informal communication with those under your supervision. And seek them out occasionally. Do not make your employees always come to you.

Nonverbal Communication

All forms of communication discussed thus far, including unofficial communication, have primarily involved the use of words. However, researchers have determined that we all constantly communicate nonverbally—with our facial expressions, our arms, our legs, our entire body—and that the nonverbal messages we transmit are more powerful in communicating emotions or feelings than are our verbal messages. An understanding of how we communicate nonverbally will help us to improve our overall communication skills. We should certainly learn enough about our nonverbal communication processes to make certain that our nonverbal messages agreee with our verbal ones.

The following incidents demonstrate the significance of nonverbal communication:

Bud Swierzcek, who has been an automotive technician with Blythe Buick for eleven years, wants the currently vacant job of service manager. He schedules an interview with Wanda Blythe, owner and manager. After being kept waiting for thirty-five minutes, he is ushered into Ms. Blythe's lavish office and seated across the huge walnut desk from her. During the discussion that follows, he notices Blythe fiddling with her glasses, shuffling papers, and gazing thoughtfully out the window. She never does seem to meet his eyes. Even though Blythe asks all the appropriate questions, Bud is discouraged. When Blythe says she will let Bud know about the job in a few days, Bud is certain that someone else will be picked. Someone else is picked.

Gerald Garapie is invited to attend a national meeting of dental hygienists. Being an ambitious, dedicated young hygienist himself, he is pleased to attend. The conference turns into a disaster for Gerry. He tries to be friendly to everyone he meets, but receives nothing more than polite brush-offs in return. A co-worker also attending the conference, Betty Watt, finally explains to Gerry that he "comes on too strong," that he intimidates people. His handshake is more than brisk; it is boncrushing. He claps people on the back and pokes them in the ribs. He looks everyone squarely in the eye and literally stares them down. Most disconcerting to strangers, according to Betty, is Gerry's habit of standing too close, almost touching the person he is talking to. "Ease up a bit," she says. "There's nothing wrong with what you say, but you scare people."

Space limitations do not permit us to study nonverbal communication in depth; however, a brief examination of some of its key aspects should be useful.

Territoriality ● Just as lions, coyotes, and even birds have their own feeding and bedding territories which they will fight to protect, so do human beings have their own individual territories. Each of us has a personal territory, a distance at which we feel comfortable in communicating with others. The exact distance varies among different cultures, with Latins and Middle Easterners, for instance, standing much closer during casual conversation than do Americans or Nothern Europeans. Additionally, each of us has several comfort distances, depending on whom we are talking with under what circumstances.

Observe the people around you as they converse. Notice how a husband and wife or a girl and boyfriend stand closer together than do less familiar members of opposite sexes. Notice too how standing close seems to indicate warmth or intimacy, while standing unusually far apart conveys coldness, aloofness, or excessive formality. Become sensitive to people's need for territory; recognize that intruding too close can be threatening, while backing too far away can indicate unfriendliness or distrust.

Territoriality is also significant in furniture and seating arrangements. The executive who sits high in a large, comfortable chair behind a massive desk is communicating nonverbally to the underling who must sit directly across the desk in an uncomfortable, straight-backed chair. An equally strong but much more positive message is sent by the interviewer who seats an interviewee in a comfortable chair beside the desk.

Posture ● The way we sit, stand, walk, or even hold our arms communicates strongly. The person who faces us directly with legs uncrossed is indicating openness; moving closer to us, back turned toward the rest of the group, is another positive signal. Crossing the legs, folding the arms across the chest, and turning the body to the side indicate uninterest or disagreement. Rubbing the nose, scratching the ear, and kicking cross legs are other negative signals.

Watch how people stand, sit, and move around while conversing. Notice how their body movements generally cohere with their spoken messages, often revealing their real intent before it is actually spoken.

Eye Contact ● We look at people's faces more than any other part of their bodies. We are all familiar with certain stereotyped facial expressions such as the cold stare, the raised eyebrow, the suspicious squint, even the come-hither look. More important than these is the way we use our eyes in normal conversation. A great deal of eye contact between two people is generally positive, but intense eye contact between two people often indicates a disagreement. In fact, two people often seem to be staring one another down in a battle of wills, with the first one to look away somehow admitting defeat. Overdone, exaggerated eye contact is often considered a sign of hostility or overaggressiveness. A low-status person staring at a supervisor is often indicating defiance. When the supervisor stares, it may indicate condescension. And excessive eye contact frequently indicates only pretended interest or concern. Any experienced teacher, for instance, has learned to recognize the seemingly attentive stare of the student whose mind is somewhere else.

Avoiding direct eye contact can mean any one of several things. It may simply indicate bashfulness or timidity. It may indicate feelings of guilt or fear. Or it may signify discomfort or even distaste.

Just as eye expressions can be deceptive, so can other forms of facial expression. A frown does not always indicate anger or distaste; it may indicate puzzlement, confusion, or deep concentration. A smile may be a sincere expression of joy, or an insincere gesture designed to ward off further contact without seriously offending. Many people hide behind seemingly permanent smiles, which are often contradicted by their other physical messages.

Hand Gestures ● Many hand gestures are familiar to us all: the pointed finger, the clenched fist, the thumbs down, the upraised palm. Many others, though, are not so obvious. Rubbing the ear or touching the nose or eye may often indicate doubt or hesitation. But pulling on the ear or rubbing the face often indicates nervousness. Fidgeting, tugging at the clothes, and tapping the fingers are other common signals of tension.

While pulling and tugging clothes may be a sign of nervousness, straightening, adjusting, or smoothing clothing are often what are called "preening gestures." Such gestures call attention to our bodies and are generally intended to attract the opposite sex. Other common preening gestures are smoothing the hair and crossing the legs, accompanied by appropriate facial expressions.

One of the most used, and abused, hand gestures is the handshake. We have all heard about the poor impression made by the wet-fish handshake. However, not everyone likes to shake hands. Others, such as many politicians and salesmen, overdo it by shaking every hand in sight.

Hand gestures often involve objects. Fiddling with a pipe, a cigarette lighter, or a pair of glasses may indicate nervousness or boredom. Methodically removing the glasses or pulling out a pipe and tobacco is often an indication of uncertainty or lack of understanding.

None of the generalizations just made about nonverbal communication is always true. No one gesture, posture, or expression always means the same. In fact, you will occasionally notice seemingly contradictory nonverbal messages. More frequently you will notice nonverbal messages seeming to contradict the verbal messages being sent. When we are insincere, our nonverbal behavior will usually indicate our true attitudes. The technical or professional person who is sensitive both to his or her own nonverbal communication and to that of others, who learns to use nonverbal communication to enhance use of words, will certainly be a more effective communicator.

Effective Listening

Of the four basic elements of communication—reading, writing, speaking, and listening—listening is the most frequently engaged in; yet it is the only one in which most of us receive no formal training. We are told to pay attention, to take notes, or to be quiet and let someone else talk, but we receive no real instruction in how to listen effectively.

This neglect of listening techniques shows. Students perform poorly on tests because they cannot listen effectively. Jobs are done improperly because a worker does not listen effectively to the boss's instructions. Contracts are lost because a salesman misunderstands a telephone order. Consider, for instance, the following example (based on an actual happening):

J. R. Walkup, Executive Vice-President of Tower Electric, is informed by mail that auditors from DeVaco, Inc., parent company of Tower Electric, need a written status report of the Thomason Furniture job within two weeks. He calls J. O. Lockman, operations manager, to instruct him to have the report prepared within a week. Lockman, who is busy trying to clear his desk of work so that he can leave on a three-week vacation to South America, responds to the vice-president's request with a "Sure, of course, I'll get right on it." But he does not bother to make a note of it or to start preparation immediately. Later that afternoon he remembers the call but, being preoccupied with other matters, confuses the Thomason and Robertson jobs. So he calls Paulette Pennington, superintendent of the almost completed Scott Robertson Lingerie Plant job, and demands that she submit a complete status report to Vice-President Walkup within the week. Ms. Pennington protests that she just submitted such a report the previous week. Still distracted by his plans of vacation, Lockman dismisses her protests and hangs up angrily. Pennington meanwhile pulls three people off other work to do the new report, thus delaying completion of the job by a full day.

When Pennington takes the report to Walkup on Friday, he is dumbfounded. Lockman is by now safely out of range in Buenos Aires, so Walkup blows his stack at Pennington, and orders her to see that the Thomason report is on his desk by Monday evening. Pennington then calls Darwin Hild, superintendent of the Thomason job, but he is so angered at the sudden loss of his weekend that he misses Ms. Pennington's explanation that the status report is for the parent company auditors. Thus he fails to include some of the financial data they need.

Finally, the third report written meets the needs of the situation. But in the process Walkup has antagonized the auditors and Lockman has almost lost his job and has made enemies of Pennington and Hild. Hild has lost company esteem. Everyone concerned has put in extra work. Two major jobs were delayed, and many dollars were spent on unnecessary overtime. All because two men did not listen effectively.

This is not an exaggeration. Such occurrences do happen regularly.

Specific Listening Problems

Distractions ● Many times we do not understand what we listen to because we either create or fail to adjust to distractions. Background noise such as machinery operating, typewriters clanging, or other people chatting should be physically shut out if possible. If not, you must learn to shut them out mentally. Also avoid creating your own distractions, such as shuffling papers, tapping your feet, humming softly, or otherwise distracting yourself, the speaker, or other listeners.

Speaker's Use of Words ● We often forget to consider the speaker's ideas because we are hung up on the words. The speaker may use highly colored words or derogatory terms such as *redneck* or *hippie* or *commie* that stir and arouse us. Do not let the words goad you into a gut-level reaction before you consider the content.

Similarly, you must learn to ignore grammatical mistakes or the use of inappropriate words at least until you have had a chance to consider the speaker's content. Also be careful about forming a positive impression merely because the speaker uses words impressively. Always look for the message behind the words.

Reactions to Speaker ● Notorious liars sometimes speak the truth; our arch enemies sometimes agree with us. We must always listen and react to what the speaker says, not to our preconceived notions of the person. Always listen with an open mind.

Uninteresting Topics ● If you expect to be bored by a speaker, you will be. Try to be interested. No subject need be automatically boring; give the speaker and yourself a chance.

Concentrating on Facts ● Listen first for a speaker's main points, then look for supporting facts and data. Many of us, especially in the classroom, are so concerned with getting the facts that we miss the general ideas that give the facts significance.

Focusing on One Aspect ● Listen to the entire presentation carefully. Do not get carried away trying to conjure up a rebuttal to one point until you have heard the speaker's other points. Be careful not to let your mind wander conjecturing about a specific point you like; again, listen to the whole.

Listening Too Hard ● Concentrate, but try not to overdo it. Trying too hard to concentrate is much like trying too hard to go to sleep: Sometimes you are better off just to relax and let it happen. Remember too that your mind works much faster than any speaker's vocal chords. You cannot help but think of something other than his or her words.

Some Constructive Suggestions

1. Do some homework. Before listening to an important speaker or engaging in an important interview or discussion, learn something about the subjects to be discussed. A speaker's comments will be clearer if you have some background knowledge of the subject.

2. Look for main points and try to get a feel for the overall structure of the talk. Build a context within which you can better understand and remember specific points.

3. Do not race ahead of the speaker to furnish the next point. You will often miss important details this way, and you may also leave thinking that the speaker actually said something that you merely imagined.

4. Do not expect to grasp fully every single thing a speaker says. Between his or her limitations as a speaker, your limitations as a listener, and the complexity of the subject, complete understanding is often impossible. If you do not understand one point, go on with the speaker to the next. Otherwise you can miss two or three understandable points while you are busy trying to figure out an earlier one.

5. Try to put aside your cares of the moment. Concentrate on the speaker, and worry about your headache, your golf game, or your family problems later.

6. Listen "between the lines" to the speaker's nonverbal messages. Listen to the tone of voice; notice the connotation of words. You can usually get a feel for the speaker's true feelings and purpose if you work at it.

7. Make mental notes of questions to ask or points to remember. Better yet, take written notes; but use some judgment. Avoid concentrating so hard on getting everything down that you understand nothing (this often happens to students in lecture courses). Make your notes thorough enough that you can understand them later. A word or two that reminds you of an entire point now will not necessarily do the same a few days, weeks, or months from now. Concentrate on main points and on figures and statistics.

8. If outlines or abstracts of a talk are distributed in advance, use them to help you to organize the speaker's words. Do not merely relax and let your mind wander, thinking you can always fall back on the outline.

9. Use the extra time your mind has between words to your advantage. Consider the types, relevancy, and validity of the speaker's evidence. Fit each remark into context. Consider unspoken implications and verbal clues such as repetition or emphatic phrasing. Review what has been said. But do not wander off the subject.

10. Minimize physical distractions. Try to get a comfortable chair and room enough to stretch out. Do not face directly out a window or directly toward a bulletin board, painting, or other distracting scene. Ignore those who want to whisper or make other forms of static.

11. Give the speaker the benefit of any doubt. Assume that he or she knows the subject and has honorable intentions.

12. Remember that you are constantly giving the speaker feedback. The experienced speaker looks for feedback and adjusts accordingly; the inexperienced speaker may be oblivious to it or may notice negative feedback and have even more difficulty completing an effective presentation. If the speaker indicates willingness to be interrupted with questions, do not hesitate to ask them. Otherwise wait. Try to respond as you would want a listener to respond to you. Look attentive, but avoid the insincere stare of feigned attention. Do not sit woodenly, but try not to distract everyone with excessive movement. Generally, show good taste and good manners.

13. Improve your vocabulary. Weak vocabulary is as much a hindrance to effective listening as it is to effective reading. Overall vocabulary development is always good, but also important is the development of the specialized vocabulary of your field. Learn the words that speakers in your profession are most likely to use. Once you get the hang of it, you will find attentive listening will itself be a big aid to vocabulary development.

14. Practice listening. Just as good writing, speaking, and reading require practice, so does good listening. And your listening ability is important enough to your total professional skills that you can certainly afford time to practice a bit. Start with the listening exercises below; then work on your own.

● EXERCISES

1. Select a partner among your classmates and conduct a simulated long-distance call. Determine roles; the receiver may be a supplier, a boss, a customer, whatever you select. The caller should do the necessary preplanning for the call, and conduct it in front of the class. Now reverse roles and conduct another call.

2. Your instructor will help you obtain four short tapes. Listen to each as follows:

    a. Take notes on this highly technical material; then take the ten-question quiz your instructor gives.

    b. Listen but do not take notes. Take the ten-question quiz your instructor gives.

    c. Outline the talk you hear. Compare your outline with those of your classmates.

    d. Listen carefully to this highly persuasive speech. Try to summarize the points made by the speaker.

14

The Job Package

Overview *Preparing an effective resume and letters of application is one of the most important tasks of your career. Develop the resume to show your credentials in the best possible light, concentrating on your professional skills and your educational background. Aim your letters of application at specific organizations, showing how you uniquely qualify for specific positions. Prepare carefully for a job interview by learning as much as you can about the position and the organization and by determining the questions you will need to ask. During the interview, follow the interviewer's lead, but be sure to stress your credentials and ask your questions.*

You cannot apply the on-the-job skills presented in the first thirteen chapters of this book until you have a job. This chapter should help you to get one. First, you will learn how to prepare an effective resume and letter of application. Then you'll learn some techniques for job interviews. Sample letters accepting and rejecting a job are included.

Resumes and Letters of Application

It is difficult to conceive of anything the average technical person· could write that would be more important to him or her than the resume and accompanying letter of application. When you write them you are playing for high stakes—a job. If they are well done, you will probably be invited to interview; if they are badly done, you will have to keep looking. Furthermore, the resume and letter of application are pieces of communication in which the facts do not "speak for themselves." Your qualifications for a given job may be good, but you still have to express them to their best advantage. There will probably be other applicants with good qualifications too. So make sure that the reader notices yours.

What the Resume Is

A resume is a clear, coherent presentation of your credentials. Although it appears to be straightforward, the resume is not the totally objective document that such a definition implies. It is photographic in the sense that it is a basic summary of your educational and occupational background, but it is also impressionistic in that it emphasizes the positive side of your character and accomplishments.

A common misconception about resumes is that they must be individualized for each prospective employer to whom they are sent. This is usually unnecessary because you can individualize the accompanying letter of application. If you are applying to several similar organizations for similar positions, you can use copies of the same resume. Many people, in fact, use the same basic resume over and over, updating it by adding entries about new accomplishments and making the necessary changes. For this reason, you should invest a great deal of time putting together a resume and making it as effective and flexible as you can. As a result, you will be able to use it in a variety of employment situations.

Resume is the most common term for a summary of credentials. But you can also use the terms *vita, vita sheet,* and *data sheet,* which mean essentially the same thing. Call yours a resume unless you are applying to someone who has specified another term. Then merely change the name atop your resume and use it as is.

What Goes into the Resume

If you were to read thirty books on the subject, you would find significant differences in what each recommends you include in the resume. This section looks at the ingredients that most authorities suggest. Some of the components are required, some are recommended for use at your discretion, and some are not recommended. Consider them all and make decisions based on your own circumstances.

1. Personal Data ● At one time it was necessary that personal information be included in the resume. Because the United States Supreme Court has clearly established that persons must not be discriminated against because of age, race, sex, or religious preference, this information is no longer required. If you wish to include personal data, such entries as age, marital status, number of children, general health, military status, and club and community organization memberships are the sort usually used.

2. Educational Background ● Your most important section may be the one giving your educational background. If you have little job experience directly relevant to the position for which you are applying, chances are good that your educational preparation is your strongest credential. Be sure to take full advantage. Study your college and high-school background for all its strong points. Consider also service schools, correspondence courses, and anything else that could be considered post-high-school education. Do not include on-the-job training; but if your company required you to attend a two-week seminar or a series of evening lectures, for example, include this kind of data.

Look especially for features of your education that distinguish it from that of other people. For instance, if most candidates for a particular job have the same academic degree that you have, look for special features in your education. You may have taken more courses in your major than were required. You may have done a significant independent study project. You may have a strong, relevant minor, or you may have participated in relevant extracurricular activities. If you think hard enough, you can usually come up with some distinctive features.

Begin your list with the college you are currently attending or have most recently attended. Then work your way back as far as the high school from which you graduated. If you have been out of high school more than ten years, you need not go into detail about your high-school background. Merely give the school, its address, and your date of graduation. If you graduated more recently, develop a full entry just as you would for the colleges you have attended.

Including the following information should give you a strong education section:

Dates of attendance.
Name of school.
Address (city and state will usually do).
Degrees or certificates.
Major and minor.
Grades (optional but recommended; if you list grades, state them in the most positive way; for instance, C+ sounds better than 2.1 of 4.0, upper one-third of class sounds better than C+).
Achievements (honors, awards, membership or positions held in organizations).
Anything else impressive you can think of.

3.   Work Experience ● Inexperienced applicants must rely on their educational background; experienced applicants can also stress their work experience. This section of the resume serves mainly as a brief summary of the jobs you have held, so do not go into detail about specific duties, achievements, and skills; you can list them later under "Professional skills." Begin with your present job and work your way back at least ten years or to your graduation from high school, listing all *significant* full- or part-time jobs. Avoid giving a long list of irrelevant high-school jobs such as lawn-mowing or baby-sitting. Include the following information about each job:

Dates of employment (month and year).
Name of company or organization.
Address (city and state will usually be enough).
Supervisor (list the name of the one who would say the most positive things about you).
Job description (a title is sufficient; if it needs explaining, one sentence is enough).

4.   Professional Skills ● If you are applying for a position in a field in which you already have some relevant experience, the section listing professional skills will be your most important. The work experience section lists all or most of your previous jobs; your professional skills section focuses on the specific kinds of skills you have acquired within your field of specialization. Organize and compose it carefully. Think back over your relevant experience, looking particularly for signs of accomplishment or for experience which others in similar jobs might not have had. If you are in data processing, for example, you can list the types of hardware and software you have used. If you are a respiratory therapist, you can list the types of machines you have used and the kinds of treatments you have administered. Because no two persons' work experience is exactly the same, however, you will have to include those specific skills that might set you apart from others in your profession. Here are some general suggestions:

Machines or equipment used.
Types of procedures used or supervised.
Special techniques learned or used.
Innovations made.
Supervisory positions handled.
Unusually rapid promotions.
Awards or other special recognition.
In some fields, types of positions held (for example, a nurse would indicate that she had two years of experience in the operating room, two years in intensive care, and one year on a medical-surgical floor; a police officer would indicate two years spent on uniform patrol, six months on traffic control, and two years in criminal investigation).
Special certification or on-the-job training (for example, certified polygrapher, registered therapist).
Experience working with very well-known people in the field.

You need not include this section if you have little or no previous work experience in your profession. But if you have acquired specific job skills, this section is essential in creating a successful resume.

5. Related Skills ● A section listing job-related skills is optional; it will be helpful to some candidates but unnecessary to others. If you have skills or abilities that might be helpful but that do not fit into your educational or professional skills section, list them here. Remember, however, to list only skills relevant to your occupational area. For instance, being a registered scuba diver would be very useful to someone in fire-fighting or police work but would be of little obvious value to a draftsman. So consider your hobbies, any armed-service experience, and your past jobs to determine if you have related skills worth listing. Here are some of the many possibilities:

Speaking foreign languages.
Having public speaking experience.
Having computer training (for jobs in other fields).
Having typing or other stenographic skills.
Holding a Red Cross lifesaving card.

6. Community Activities ● Another optional section that may be quite helpful or totally irrelevant is one that gives your community activities. Many positions call for a great deal of direct contact with the public, while others require you to know influential public figures. Also, many companies like their employees to be involved in the community. Consider your background and the field you are going into. If you can come up with a few activities that might be helpful, list them. Some typical activities are:

Service clubs such as Rotary, Kiwanis.
Religious groups.
Boy Scouts, YMCA, and the like.
Charities.
Volunteer groups such as Candy Stripers.
Citizens' advisory groups.

7.   Hobbies and Personal Interests ● Although some people include hobbies and personal interests in their resumes, there is little evidence to indicate that such information on a resume is of any benefit. The only hobbies to list are those that qualify as related skills.

8.   Salary Desired ● The basic argument in favor of including a section specifying salary desired (or minimum salary acceptable) is that you can inform a prospective employer of your salary expectations before any interview is set up. The arguments against including such a section are that suggesting too high a figure could prevent you from being offered a job with a slightly lower starting salary but with excellent chances for advancement and fringe benefits and that suggesting a specific figure could prevent you from being offered a higher salary. Therefore, many candidates do not mention a specific salary on their resumes, either relying on their knowledge of the job market to help them avoid applying for low-paying positions or waiting until the interview in order to discuss the exact figure.

Try to avoid mentioning a salary, but if you do, use the term *salary desired* and give the amount you would like rather than a minimum figure.

9.   Position Desired ● Specifying the position you want is another debatable practice. Many authorities consider it absolutely essential; others do not recommend it. The real problem here is that it is difficult to phrase a description of the position you desire without doing some harm. Many such descriptions give no concrete information and usually sound a little pompous. Typical is "Position desired: An opportunity to apply my electronics and supervisory experience in a modern, progressive, growth organization." More specific descriptions pose a different problem in that they restrict your opportunities. For instance, "Position desired: Junior mechanical draftsman" would limit you to positions at that level when you might qualify for a higher position. Furthermore, if your job description is too specific, companies using different terminology may have difficulty determining what kind of position you have in mind.

A reasonable compromise is to state clearly in your accompanying letter of application exactly what position you have in mind but not to mention it in your resume. However, you should include the section in the resume when you think that someone might misread your intentions; for example, if you were changing professions or if you had substantial experience in two different fields, a job description specifying the field in which you sought employment would be helpful.

10.   References ● Authoritative opinion as to the inclusion of references in resumes differs sharply. Some people do not mention references on their resumes. Others state that they are available upon request. Including them seems best. It is true that many employers do not take references seriously and that even the weakest job candidate can usually come up with favorable references. But some employers do take references seriously, and others insist on having them as a matter of protocol. While the only real damage done by including them is to lengthen the resume, not including them might have more harmful consequences.

Try to get three or four people to agree to furnish references. If you have included names of former supervisors in the work experience section, you need not list them again as references. If you have not mentioned them before, list one or two here. If you have had little or no relevant experience working under supervisors who know your work well, try to get variety in your references. The names of one or two former instructors, counselors, or deans would be helpful. But avoid mentioning three or four of your former professors; their impartiality might be suspect, and their knowledge of you is limited. Family friends can provide excellent character references, but try to select those whose professional position, such as general manager, vice-president, M.D., or attorney, for example, would lend authority to their evaluation. You also might want to list any clergyman you know well.

Be certain to ask permission before using anyone's name. If a person is reluctant to agree, try someone else. Make sure that the people listed will give you strongly positive recommendations. Include with their names their job titles and business addresses.

11.   Availability ● Tell the prospective employer when you could begin work. Tell him how much notice prior to your termination date you must give your present employer. Some people also include their feelings about relocating. Typical statements are "I do not wish to relocate" or "I am willing to relocate anywhere in Northern California, Washington, or Oregon," or "I am willing to relocate anywhere in the United States."

Putting the Resume Together

When you have gathered all your information and have decided what to include, your next step is to work up a draft. Begin by deciding in what order you will arrange the sections. Although practice varies, there are some basic guidelines.

1. If you ask for a particular position, put it first. Everything else in the resume will be judged against it.

2. If you include a personal data section, either put it first (after the position desired, if that is included) or near the end. Personal data is obviously different from work and educational experience or professional skills, so keep it separate.

3. References (if you include them) and availability normally go last.

4. Here are two possible arrangements based upon these suggestions. Depending on your background, you should plan to include most of the categories below.

I	II
Personal data	Position desired
Educational background	Salary desired
Work experience	Professional skills
Professional skills	Work experience
Related skills	Educational background
References	Community activities
Availability	Personal data
	References
	Availability

Polishing the Completed Resume

Make your resume mechanically perfect, factually accurate, and visually attractive. Many prospective employers admit that the form and appearance of a resume are important factors in their evaluation of a candidate. They believe that if an applicant cannot take the time and effort to produce a first-rate resume, he or she might not have the right kind of work habits. Whether this is good reasoning is not important; what is important is that it happens.

Refer to Part 3 of this book to check your mechanics, and use a dictionary to check your spelling. Type your final draft on a typewriter with a good ribbon and clean keys and on high-quality paper. If you send out several resumes, arrange for a printshop to make copies by using an offset process. Do not send out Xerox or carbon copies. Here is a checklist to use in getting your resume ready for mailing.

1. Use correct grammar, but you need not write complete sentences. In striving for brevity, you can use short clauses or phrases.

2. Use abbreviations to save space (names of states, well-known organizations and clubs, and the like).

3. Single-space for a compact appearance.

4. Do not omit any worthwhile qualifications merely for the sake of brevity.

5. Use reverse chronological order for lists of schools attended and jobs held.

6. Include all pertinent dates; providing the month and year is sufficient. Try not to leave gaps of more than two or three months.
7. Include at least city and state for schools attended and previous employers.
8. Do not volunteer damaging information unless you want to let your reader know about a negative facet of your background that you think will come up during the interview.
9. Put your facts in the most positive and favorable way you can, but do not distort the truth. By all means, *do not lie.*
10. For references, try to include the names of influential people who will speak favorably of you.
11. Double-check each section to make sure that you have not forgotten any useful entries.
12. Make the final copy clean and neat, even if you have to have it typed by a professional typist.
13. Put your name, address, and phone number at the top.

The two sample resumes in Figures 14.1 and 14.2 illustrate the two basic types outlined earlier. The first is from a person who has had no relevant work experience, while the second is from someone who is emphasizing extensive job experience and skills.

Writing the Letter of Application

Each resume sent to a prospective employer should be accompanied by a letter of application. The resume presents all your credentials; the letter adapts specific credentials to a specific position. A good application letter usually contains the four following paragraphs:

Paragraph 1 identifies your source of information about the position and states your interest in it.
Paragraph 2 mentions one or, preferably, two key features of your background that are relevant to the position.
Paragraph 3 indicates an interest in the job, organization, or geographical area.
Paragraph 4 suggests a possible time for an interview.

Here is a checklist to consider when putting the letter together.

1. If you do not know the name of a specific person to whom the letter can be addressed, send it to the personnel director of the organization.
2. If you are responding to an advertisement or placement-service listing, make sure that you specify the job you are interested in.
3. Paragraph 2 is the key; list at least one feature in your background that other candidates for the job could not likely match. The feature

```
                      Resume of Alice M. Scorby
            2162 Country Club Rd., Orlando, FL   32807
                         314-297-4166


                          Personal Data

Age:  20        Health:  Excellent          Ht.:  5'7"
                Marital Status:  Single      Wt.:  127 lbs.


                     Educational Background

Sept. 1979-April 1981   Valencia Community College, Orlando, FL.
   A.S. in Registered Nursing.  B+ average in nursing courses.
   State Board examination passed, July, 1981.  Active in SNAF
   and AGD society.

Sept. 1975-June 1979   Edgewater High School, Orlando, FL.
   Graduated in upper 1/4 of class.  Member paramedical club,
   student council, and science club.  Varsity cheerleader.


                        Work Experience

January 1976-present   McDonald's, 1800 S. Semoran, Winter Park, FL.
Counter girl, 20 hours weekly during school, full-time during vaca-
tions.  Supervisor, Walter R. Reade.


                          References

Dr. Theodore Dippy            Mr. Louis Folguerus
Edinburgh Drive               Director of Health Programs
Winter Park, FL  32791        Valencia Community College
                              1800 Kirkman Road
Mr. Ralph Macon, Head Cashier Orlando, FL  32802
First National Bank of Apopka
Apopka, FL  32701


                         Availability

I am ready to begin with one week's notice.  I will consider
relocating to any major Southeastern city.
```

Figure 14.1 Sample Resume: No Relevant Experience

Resume of
Marvin P. Barkum
177 Center Street, Atlanta, GA 27931
616-555-8181

Position desired: Senior systems analyst in a large data center
    with opportunity for advancement to supervisory positions.

Salary desired: $21,000

Professional skills: Eight years' data processing experience at
    all levels from key punch through systems analysis. Three
    years as analyst for large utility firm. Experienced in
    software of all types; specialities in accounts receivable
    and payroll. Working knowledge of Fortran, Cobol, Neat-3,
    PL-1, and Basic. Experienced with entire IBM and NCR lines.
    Have regularly received merit raises and commendatory
    appraisals.

Work experience:
    September 1971-present North Georgia Utilities, Atlanta, Ga.
    Computer operator and systems analyst. Max Weilop, supervisor.
    (You may contact Mr. Weilop if you wish.)

    October 1969-July 1971 Atlanta Park Board. Recreations
    director, Angela Dubini, supervisor.

Educational background:
    September 1977-present North Georgia Junior College, Atlanta.
    Thirty-nine hours in data processing courses with A- average.

    September 1969-June 1971 DeKalb Community College, Stone
    Mountain, Georgia.
    A.S. Degree in Business Management, 3.4 of 4.0 average.
    President of DECA chapter, several DECA awards.

    July 1966 Graduated from Crispus Attucks High School,
    Indianapolis, Indiana.

Community activities: Member Jaycee and Kiwanis clubs. Volunteer
    worker in Neighborhood Youth Corps and Big Brothers of
    America. Scoutmaster. Active in many activities of
    Tabernacle Baptist Church.

Figure 14.2 Sample Resume: Extensive Work Experience

```
                    Resume of Marvin P. Barkum, page 2


  Personal data:  Age 32, married, 3 children, 6'1", 165 lbs.,
        health excellent.
        Military:  USMC 1966-1969, honorable discharge.

  References:  Ms. Gayle Goodin, Instructor of English, DeKalb
        Community College, Stone Mountain, Georgia 30088
        Mr. Robert Pew, Data Processing Director, North Georgia
        Junior College, 1800 Ivanhoe Blvd., Atlanta, Georgia 30327
        The Reverend Lucas Hartman, Minister, Tabernacle Baptist
        Church, 1780 Claire de Lune Way, Atlanta, Georgia 30315
```

Figure 14.2 (Continued)

might be a related skill, a strong academic background, exceptionally strong references.

4.   When pointing out your key feature, work in a mention of the enclosed resume. Your real goal here is to get the reader to look at the resume.

5.   Since no employer likes to feel that he or she is just one of many to whom you are writing, find out something about the specific position and organization. Let Shell Oil know, for example, that you are sincerely interested in Shell, not merely in any oil company.

6.   Your willingness to have an interview at the employer's convenience, but also indicate when an interview would be most convenient for you.

7.   Avoid pat and insincere endings such as "thanking you in advance" or "humbly grateful."

8.   Try to give the letter a brisk, positive, businesslike tone. Be confident but not egotistical. Try to limit your letter to one typewritten page. Do not recite too many features; your resume has them all.

9.   If you are responding to an advertisement or a direct request, study the ad or request to help you determine which accomplishments to stress.

10.   Feel free to send letters and resumes to companies who regularly hire people with your qualifications, even if you are unaware of any immediate opening.

11.   Work just as diligently at making your letter mechanically sound and attractive as you do with your resume. Candidates with good resumes have been rejected because of poorly constructed letters.

The letters of application shown in Figures 14.3 and 14.4 were written to accompany the two sample resumes. Figures 14.5 and 14.6 are sample responses to a job offer.

2162 Country Club Road
Orlando, Florida 32807
October 18, 1981

Director of Nursing
Raleigh General Hospital
Raleigh, North Carolina 27180

Dear Director:

Would you please consider my qualifications for a position as a Registered Nurse on your staff. I am interested in a current opening; but if none are available, I would appreciate this letter of application being placed on file for future consideration.

As the enclosed resume shows, I recently received my Associate Degree from Valencia Community College with high marks in my nursing courses. Mr. Louis Folguerus, Valencia's Director of Allied Health Programs, can provide a strong recommendation based on my academic work. My interest in nursing goes back to paramedical club in high school, and I am now anxious to apply my training and interests to a nursing career.

Several of my nursing instructors have recommended Raleigh General as a good, large, innovative hospital where I could learn the newest and best nursing techniques. Working at Raleigh General would also enable me to continue my education by taking advanced nursing courses at one of the area universities.

My family and I will be spending the first two weeks of November in Durham, and I would like very much to come to Raleigh for an interview during that time. If my credentials are of interest to you, please let me know when you would like me to come to Raleigh.

Sincerely yours,

Alice Scorby

Alice Scorby

Enclosure

Figure 14.3 Letter of Application

177 Center Street
Atlanta, Georgia 27931
October 14, 1980

Mr. Richard McGhee
Personnel Director
W. A. Ward Company
420 Peachtree Street
Atlanta, Georgia 27907

Dear Mr. McGhee:

Your advertisement in the October 11 <u>Constitution</u> for a
senior systems analyst is just the opening I have been looking
for. Please consider my qualifications for the position.

As you will note on the enclosed resume, my nine years of
programming and systems experience at North Georgia Utilities has
familiarized me with all the hardware and software you use. For
the past two years, I have worked a great deal on the IBM 370/15,
your main computer.

Many of my acquaintances at W. A. Ward and at other area
data centers have spoken very enthusiastically about your sound
personnel practices, especially your willingness to promote employees who work harder and more efficiently than their colleagues.
North Georgia is an excellent organization, but I feel I could
better demonstrate my capabilities at W. A. Ward.

If you are interested in my qualifications, I would be happy
to discuss them further with you. With a few days' notice, I
could come in any workday afternoon for an interview.

Sincerely yours,

Marvin P. Barkum

Marvin P. Barkum

Enc.

Figure 14.4 Letter of Application for Advertised Position

1431C San Juan Terrace
Columbus, Ohio 43137
November 16, 1981

Mr. Robert C. Ross
Chief of Research
Vane Chemical Company
Akron, Ohio 42133

Dear Mr. Ross:

Since I am eager to begin using my training in a well-equipped department such as yours, I am happy to accept your salary offer of $1,100 a month. Thank you for your confidence in my abilities.

Your suggested starting date of January 3 is ideal. I will be moving to Akron between Christmas and New Year's and will report to your office on January 3. If there is anything else I need to know before reporting, please write me at the address above.

Sincerely yours,

Sara C. Raymond

Sara C. Raymond

Figure 14.5 Letter Accepting a Job Offer

1431C San Juan Terrace
Columbus, Ohio 43137
November 16, 1981

Mr. Robert C. Ross
Chief of Research
Vane Chemical Company
Akron, Ohio 42133

Dear Mr. Ross:

    Your invitation to join your staff at such a generous salary
is very complimentary. Your facilities are the very best I have
seen, and everyone there seems very professional.

    However, you will remember that I had earlier expressed
interest in trying to stay in Columbus. I have recently received
a good offer locally and have decided to stay here.

    Please accept my thanks and my best wishes for you and your
department.

                  Sincerely yours,

                  Sara C Raymond

                  Sara C. Raymond

Figure 14.6 Letter Turning Down a Job Offer

Job Interviews

If your resume is successful, your next step will probably be the job interview. In fact, you will likely interview with several concerns before being offered and accepting your first professional position. Then, should you change organizations, you will likely repeat the process, perhaps several times during your career.

And the way you conduct yourself during these interviews will often determine whether you are offered the job. Your resume can be a work of art, but it will merely get you the invitation to interview. If the job is desirable there will be other candidates whose resumes also look impressive and who will also be invited to interview. Often five or six candidates with equally impressive credentials will be called, with the interview alone determining which one is offered the position.

Fortunately, you will not have to go into that initial interview unarmed, strictly at the mercy of some brutal interviewer who will try to dig out your hidden flaws. In the first place, interviewers are human, and most of them are more concerned with finding your strengths than with making you look bad. Secondly, there are specific ways to arm yourself to make certain that you do a good job.

Preliminary Preparation

Getting ready for a job interview begins long before you put on your best dress or the new business suit you have bought for the occasion. Direct preparation should begin almost as soon as an interview time and place are confirmed. Indeed, you should have been getting ready for interviews from the time you began seeking a job.

Most of the preparation for a job interview can loosely be termed "boning up." Two areas need attention: you, and the organization you will interview with. Since you already know quite a bit about yourself, especially since you have recently prepared a resume and letters of application, begin studying the organization. Later you can focus on yourself. First, learn whatever you can about them. You need not spend hours studying financial statements or learning the names of all of their employees, but you must at least learn some basic information. You do this for several reasons: you want to know what questions to ask the interviewer (who expects you to have questions of your own as well as answers) and you want to show the interviewer you have some knowledge of the organization you are considering working for. Probably the worst impression you can give an interviewer is that you know nothing about the organization other than that it employs people in your field.

So begin by learning some basic facts about the organization. What does it do? If it is an architectural firm, what kinds of jobs does it do? Does it specialize in large condominium complexes, in office buildings, or in single-family houses? If it is a hospital, how large is it? What special facilities or departments does it have? With the specialized knowledge you have of your profession, you should be able to determine what information to seek. Here is a sample checklist of questions a candidate for a police officer's job might try to answer about a prospective department.

> Info to get about Moline City Police Department
> 1. Crime statistics from past few years?
> 2. Number of officers?
> 3. How well equipped?
> 4. How well funded and paid?
> 5. How are they organized—appointed chief? etc.

Our hypothetical candidate will add more questions to the list, but these will do for a start.

Once you have familiarized yourself with the organization and have learned how it differs from similar or competing ones, narrow your search down to the branch, department, or position you would be working in. As a respiratory therapist, now that you know something about the hospital, try to learn about its respiratory therapy department. What is its overall reputation? Is it known as innovative or does it just muddle through? How is it administered? How and how well is it funded? Is it well equipped? Much of this kind of specific information will not be readily available; you will have to wait until the interview to inquire about it. But a check with professional acquaintances, colleagues, classmates, or instructors will often turn up some useful data. And professional organizations can often help. Let us assume that the would-be police officer of the earlier example was interested primarily in working with juveniles. He might add the following to his list:

> 6. Separate juvenile division?
> 7. Juvenile head at what rank?
> 8. How large?
> 9. Divisional reputation?
> 10. Cooperation from area courts?
> 11. Facilities such as juvenile detention center?

Now you can narrow your study to the specific position you are seeking. Remember, your purpose is still twofold: you want to learn what you can now, but you also want to determine what questions to ask of the interviewer. Check into salary, promotional policies, benefits, organizational structure, and any considerations you may have about working conditions. You will find answers to many of your questions; those you cannot answer now will be saved for later. The law-enforcement candidate might jot down the following:

> 12. Beginning patrolman's salary?
> 13. Probationary period?
> 14. Promotional policies?

15. Benefits?
16. Union? Fraternal association?

With a clear notion of what kind of organization you are considering, of the division you would be entering, and of the specific nature of your job within the organization, you are finally ready to reconsider your qualifications. Double-check your basic qualifications: your age, college background, job experience, height and weight if appropriate. Make sure you have at least minimal qualifications for the position. Then consider your resume and letter of application. You must have some definite strong point or you will not even be invited for the interview. Be prepared to stress this strength. Next go beyond the obvious strengths such as outstanding college preparation or relevant prior experience. Find other, less obvious strengths. Consider yourself in the job in question. What can you offer the organization? What will make you a better firefighter at the Reedy Creek Fire District than someone else who is applying? Chances are your competitors have good qualifications too. You need to show up as just a bit better. The potential juvenile officer might emphasize the following:

1. Three years shore patrol.
2. Volunteer work in four youth groups.
3. Extra courses in adolescent and child psychology.

Note that he did not bother to list his obvious basic qualifications.

Making Ready

At least a day or two before the interview, prepare some questions for the interviewer. While you are not expected to burst into the office asking all sorts of questions, you are expected to have some. As an earlier section pointed out, most interviewers want you to ask questions. Anyone not interested in working conditions, advancement, supervisory policies, and salary is not considered a desirable candidate. So take some time to decide just what you would like to know, jot down some questions, and be prepared to ask them. The hypothetical candidate for the police officer's job might ask the following:

1. How large is your juvenile division, and who heads it?
2. What juvenile facilities do you have?
3. Do you have any sort of special juvenile programs, independently or with community groups?
4. What is the probationary period for new officers?
5. What is the starting salary? Is it increased after the probationary period? What are the increments?
6. What are your promotional policies? Tests? Years in rank required?
7. What other benefits, such as insurance or uniform allowance, are offered?
8. What are the work hours? Do shifts rotate regularly?

Many candidates like to carry their lists of questions into the interview with them. If you do so, be careful not to read it to the interviewer; rather, use it for reference to make certain you do not forget anything. You can probably get along without the written list if you take the time to consider your questions before the interview.

Final Polishing

Perhaps not many job offers are lost because of a candidate's appearance, but why take chances? Even if you are applying for a job in which you will wear work clothes and get dirty or greasy, dress up a bit for the interview. Suits or dresses are typical. For a job such as rock musician or salesperson in a mod clothing store you would want to dress accordingly; otherwise dress conservatively. Extreme hairdos and the like might be matters of personal taste to you, but the exotic and the modish will alienate some interviewers. So use good judgment about your appearance.

Starting the Interview

The first impression is probably a bit overemphasized, but it is important. Walk into the room, say hello, and introduce yourself. If the interviewer offers to, shake hands, firmly but without smashing knuckles. If you are asked, sit down; otherwise, remain standing for the moment. If the interviewer wishes to make small talk, join in. If he or she gets right to business, respond in kind. Let the interviewer set the pace.

During the Interview

Discuss what the interviewer wants to discuss. Answer questions fully, but do not launch into lengthy dissertations. Explain your yes or no answers. Broad, open-ended questions such as "What is your philosophy of education?" or "Why do you want to be a supervisor?" should receive specific answers. Do not worry about formulating the perfect answer. A bit of stammering or hesitating will not hurt. Just speak honestly and specifically. If the interviewer cuts in when you seem to be rambling a bit, take the hint.

Many interviewers will furnish information about the job and the organization; others will ask if you have any questions. Otherwise, feel free to ask your questions whenever there is a convenient lull. Should the occasionally dominating interviewer make that impossible, bide your time. If no good opportunity arises and the interview appears to be ending, say that you have some questions to ask. Keep in mind throughout that, while the interviewer generally sets the tone, your positions are similar. The interviewer wants to learn about you and size you up; you need to learn about a job and a company and size them up. Most interviewers are aware of this, but if you run into one who is not, do

During your job interview, keep in mind that even though the interviewer wants to learn about you and size you up, it's just as important that you learn about a job and a company and size them up.

not hesitate to politely assert yourself. Above all, do not leave the interview with unanswered questions, unless, of course, you have decided against the position.

Ending the Interview

In ending the interview, again take your cue from the interviewer, who will usually signal by some comment such as "Well, Mr. Reed, it will be about two weeks before we decide. We'll be getting back to you then." Sometimes the cue will be a pause or a "Well, Mr. Reed, is there anything else?" If you have learned all you need to, accept the cue and leave with a polite "thank you." Do not, however, leave until you have learned all that you need to and until you know your status concerning the job. Again, most interviewers will tell you definitely when they will decide; they might even make an offer on the spot.

Should an interviewer fail to let you know when the job decision will be made or what your chances are, ask. If you are definitely interested in the job, state that you would like to know something definite soon (a week or two should suffice) so you can make plans. Or you can mention that you are considering other organizations and that a quick decision here would be helpful. If you have decided against the position, let the interviewer know. Both parties deserve to know their status before the interview breaks up.

Some Dos and Don'ts

1.   Do stress what you could do in the new position.
2.   Do mention any real strengths that may not have shown up on your resume.
3.   Do answer questions honestly and fully.
4.   Do dig a bit with follow-up questions if the interviewer hedges or avoids your questions.
5.   Don't try to hide weaknesses in your background. You are not expected to volunteer damaging information, but neither should you lie about it or gloss over it.
6.   Don't brag about your accomplishments.
7.   Don't criticize your present or past employers.
8.   Don't use sob stories; you will get the job because of worth, not because of need.

• EXERCISES

1.   Prepare a resume that you could use if you were job-hunting right now.
2.   Update the resume as you would expect it to appear in a few months or years when you receive the degree you are working toward.
3.   Prepare a resume for an acquaintance who has more than five years' relevant professional experience. If you have that much experience, prepare a resume for an acquaintance who is just starting.
4.   Write a letter of application to send along with a current resume to apply for a position advertised in a local newspaper. Clip and attach the ad.
5.   Write a letter of application that you might use upon graduating or making your next major professional advancement.
6.   Assume that you are preparing for your first interview for the position you have been working towards. Prepare a list of questions you will want to ask. Then identify at least three of your strengths to stress during the interview.
7.   Write a letter politely rejecting the job from Exercise 6.
8.   Write a letter accepting the job from Exercise 6.

PART

THREE

Handbook for Reference and Review

15

Sentence Elements and Patterns

Basic Elements of the Sentence

> Verbs
>
> Nouns
>
> Adjectives and Adverbs
>
> Pronouns
>
> Structure Words

Types of Sentences

> Common Patterns
>
> Expanded Sentences

Overview   To develop sentence sense—the ability to recognize and construct gram-matical sentences—begin by learning to identify basic elements: verbs, nouns, adjectives, adverbs, pronouns, and various structure words. Then, by joining these elements into basic patterns and expanding and combining the patterns, you can build an almost infinite variety of sentences.

The term *sentence* has been defined in many ways, none of them wholly satisfactory. Rather than memorizing one of these definitions, you will do better to learn to identify the parts of speech—the blocks used for building sentences. Then you can familiarize yourself with the basic patterns used for sentences. Recognizing the basic types of blocks used and the most common patterns for putting them together into effective sentences helps you develop what is often referred to as *sentence sense,* the ability to recognize, almost intuitively, a grammatical sentence.

Basic Elements of the Sentence

Verbs

Two groups of words, verbs and nouns, are basic to building grammatical English sentences. Without both, a group of words cannot correctly be called a sentence. Verbs have historically been defined as "words that show action or state of being," and this definition will still work.

Most verbs are what we call "action words." They do just what the term implies—they show something happening. Notice the action verbs in the following sentences:

> The lieutenant *wrote* a report on the break-in.
> A technician *assembled* the parts.
> Barry *resigned.*
> My supervisor *praised* my work.

The second group of verbs—those that show state of being—are often called "linking" or "copulative." While action verbs number in the thousands, the list of common linking verbs is quite short: forms of the verb *be* and *become, seem, appear, look, sound, feel, taste,* and *smell.* As the next section shows, linking verbs form certain sentence patterns and action verbs form others. Here are some typical sentences containing linking verbs.

> The lieutenant *is* a brilliant man.
> The parts list *was* accurate.
> My template *looks* badly worn.
> The sample *smelled* rancid.

In addition to showing action or state of being, verbs also generally show the time of action or being. For instance, in "The lieutenant *is* a brilliant man" *is* indicates the present whereas in "The lieutenant *was* a brilliant man," the past is indicated. "The lieutenant *will be* a brilliant leader" indicates the future. This phenomenon is commonly called *tense*. The tense system of English is complex and a discussion of it is beyond the scope of this book. However, learning to identify a verb tense as past, present, or future is a good beginning. Then, for an explanation of distinctions within those broad categories, see a more complete English handbook (e.g. the *Writer's Guide and Index to English*, Ebbitt & Ebbitt, Scott, Foresman, 1978). The following examples show some varieties of past, present, and future tenses:

Past:	He completed the quilt last night.
	He had completed the quilt before he went to bed.
Present:	He is completing the quilt today.
	He completes a quilt every week.
Future:	He will complete a quilt Friday.
	He will have completed sixteen quilts.

Action verbs possess another important quality—voice. An action verb may show either the active or the passive voice. The active stresses the doer of an action. The passive stresses the receiver of that action:

Active:	Ms. Thomas completed all the tests.
	Mr. Douglas wrote a preliminary report.
	Bernie had damaged the oscilloscope.
Passive:	The tests were completed (by Ms. Thomas).
	A preliminary report was written (by Mr. Douglas).
	The oscilloscope had been damaged (by Bernie).

Notice that the passive form of each sentence may or may not indicate the doer of the action. Each sentence is grammatically complete without the phrases in parentheses.

The active voice is used more than the passive, and many people consider the passive to be stylistically inferior. Business and technical writers are, in fact, often criticized for overusing the passive. The active voice does generally lead to more lively, readable prose. But feel free to use the passive anytime you wish to stress the receiver of the action rather than the doer.

Nouns

Along with verbs, nouns are essential building blocks of sentences. Nouns are names; the name given to anything, real or imagined, tangible or intangible, is a noun. The following brief list illustrates some nouns:

tree	tinsnips
woman	Colorado
chicken	honesty

Nouns perform the action shown by action verbs; it is their existence that is announced by state of being verbs. When something happens, it is done by a noun to a noun. Notice the nouns in the following sentences:

Water-piks are used to clean *plates.*
Electrolysis is our basic *technique.*
Fred Prince completed the *blueprints.*
The *tenants* made too much *noise.*

Nearly all nouns change form to indicate their number—either singular or plural:

one plant	two plants
one grass	several grasses
one mouse	three mice

They also change form to indicate possession:

the briefcase's handle
the briefcases' handles
Mr. Grace's assistant

Nouns are often easily identified by certain words that often precede them. *A, an,* or *the* can precede almost any noun, as can possessive words such as *her, my, your, supervisor's.*

a tachometer	an optometrist	the executive
her office	my idea	the tools

Adjectives and Adverbs

Verbs and nouns, as we have already noted, are the two necessary classes of English words. Two other groups, adjectives and adverbs, also help convey the content of typical sentences. Adjectives and adverbs are commonly known as *modifiers.* Adjectives modify nouns; adverbs commonly modify verbs and often modify adjectives or even other adverbs. To *modify* means to describe, qualify, or limit the meaning of another word. Notice the changes when adjectives are used to modify nouns in the following sentences:

He used a pencil for the drawings.
He used a *4-11* pencil for the *preliminary* drawings.

The technician extracted samples.
The *laboratory* technician extracted *blood* samples.
The *soil* technician extracted *four muckland* samples.

The salesman approached the customer.
The *real-estate* salesman approached the *reluctant* customer.

The following sentences demonstrate how adverbs can change the meaning of sentences by modifying verbs:

My supervisor told me to do my work.
My supervisor told me *angrily* to do my work *faster*.

Our new feeding program progressed.
Our new feeding program progressed *slowly*.
Our new feeding program progressed *rapidly*.

The next group of sample sentences shows how adverbs can modify adjectives or other adverbs:

The short notice we were given created serious problems.
The *unreasonably* short notice we were given created *extremely* serious problems.

She never reads the long forms.
She *almost* never reads the *unnecessarily* long forms.

In addition to noticing their functions in sentences, you can use several other techniques to recognize adjectives and adverbs. You can identify adjectives, for instance, by testing them in this sentence:

The _____ person is very _____.
The *tall* person is very *tall*.
The *honest* person is very *honest*.
The *intelligent* person is very *intelligent*.

Nearly all adjectives fit this pattern.

Adjectives can also be compared. By adding *er* or *est*, or in some cases *more* or *most*, you can indicate the degree to which a noun has a certain quality. Here are some examples:

short	shorter	shortest
tiresome	more tiresome	most tiresome
lazy	lazier	laziest
beautiful	more beautiful	most beautiful

Many adverbs, too, can be compared, usually by adding *more* or *most*.

recently	more recently	most recently
savagely	more savagely	most savagely
soon	sooner	soonest

Another useful clue to identifying adverbs is their *ly* ending. Some adverbs do not end in *ly*—*soon, near, often, never,* and *now,* for instance—but the vast majority do. A few adjectives, such as *lonely, timely,* and *friendly,* also end in *ly,* but again their number is few. When you see a word ending in *ly,* chances are that it is an adverb. In fact, many adverbs consist of an adjective with *ly* added. Note the following examples:

quick	quickly
slow	slowly
ready	readily
excited	excitedly

Pronouns

Pronouns are traditionally defined as "words that take the place of nouns." The noun replaced by the pronoun is referred to as the *antecedent.* Pronouns can be divided into various types; five groups are discussed here: personal, relative, demonstrative, interrogative, and indefinite.

Personal pronouns possess a characteristic called *case* that confuses many writers. The three common cases are generally termed *nominative (subjective), objective,* and *possessive.* A different pronoun form is required for each case. Here are the personal pronouns listed by case:

Nominative	Objective	Possessive
I	me	my, mine
you	you	your, yours
she	her	her, hers
he	him	his
it	it	its
we	us	our, ours
they	them	their, theirs

The appropriate case is determined by the pronoun's function in a given sentence. This is further discussed later.

Relative pronouns replace nouns, but they serve a linking function. Specifically, they join a dependent clause to the rest of a sentence. The five commonly used relative pronouns are *who* (nominative), *whom* (objective), *whose* (possessive), *which* (no case), and *that* (no case). Below are examples of sentences containing relative pronouns:

Mrs. Jones is the one *who* fired me.
Mrs. Wilson was the one *whom* I worked for.
Mr. Brown had formerly held the job *that* I now hold.
Barbara Finchitti manages the store *whose* sales increased the most.
Barry Nieman works in Room 327, *which* is right next door.

Interrogative pronouns ask questions: *who, whom, whose* and *which* are the most common:

Who won the game?
Whom will we play next week?
Which do you want?

Demonstrative pronouns demonstrate or point to a noun. *This, that, these,* and *those* are commonly used demonstrative pronouns. Notice in the examples below that the first three contain demonstratives pointing to nouns immediately following them, while *those* in the final sentence is not followed by a noun. The antecedent of *those* would have to appear in a previous sentence.

This file is the one you want.
That laboratory is off limits to visitors.
These designs are your best yet.
Those are not nearly so good.

Indefinite pronouns are one of the largest and most confusing groups. They function essentially as nouns, making only indefinite reference to persons or things. They often have no specifically stated antecedent. Following are commonly used indefinite pronouns:

one	somebody	anybody	anyone
another	everybody	one another	each other
no one	nobody	everyone	someone

Structure Words

Verbs, nouns, pronouns, adjectives, and adverbs carry the informational content of English sentences. However, this information can only be conveyed clearly and fluently when other words, often called *structure words,* are used effectively. These words comprise several overlapping and often confusing classes.

Prepositions ● A preposition joins with a noun or pronoun (called the object of the preposition) to form a prepositional phrase. The phrase functions in a sentence as a modifier, much as an adverb or adjective does. Here are some common prepositions:

above	between	near	tell	across
beyond	of	to	after	by
off	toward	among	during	under
on	at	for	over	upon
before	from	past	with	below
into	through	without	ahead of	in back of
in spite of	in place of	because of	inside of	rather than
on account of	apart from	in case of	as well as	aside from
by means of	contrary to	together with	with regard to	

Some typical prepositional phrases are demonstrated in the following sentences:

In spite of the delay, we finished *on schedule.*
Near the property line is a grove *of walnut trees.*
From January to December he was involved *in litigation.*

Conjunctions ● Two types of conjunctions—coordinating conjunctions and subordinating conjunctions—are used to connect one part of a sentence with another. Coordinating conjunctions join two words, phrases, or clauses of the same grammatical type. Subordinating conjunctions join an otherwise complete sentence (referred to as an *independent clause*) with a group of words that do not form a complete, grammatical sentence (usually termed a *subordinate* or *dependent clause*).

Coordinating Conjunctions		Subordinating Conjunctions	
and	both . . . and	more than	so that
but	either . . . or	no matter how	though
yet	neither . . . nor	provided that	unless
so	not only . . . but also	since	until
or	nor	as long as	when
		as soon as	where
		because	while
		before	in order that
		after	inasmuch as
		even though	if

The first three sample sentences below demonstrate the use of coordinating conjunctions; the last three demonstrate the use of subordinating conjunctions:

Ms. Baker *and* Mr. Browning will attend the meeting.
I have a v.o.m., *but* I need some other equipment.
Not only will we install the terminal *but* we will *also* service it for the first year.

Until we get the proposal ready, we will have to work double shifts.
Since we need so much new equipment, let's replace the whole lot of it.
He radioed in the vehicle's license number *before* he approached it.

Conjunctive Adverbs ● Another common connective that functions primarily to join two independent clauses is the conjunctive adverb. The following are those most frequently used:

accordingly	earlier	however	otherwise
afterward	furthermore	moreover	still
consequently	hence	nevertheless	therefore

In the following sentences note that the clauses joined by conjunctive adverbs could stand alone:

> Ms. Thompson had formerly worked in an operating room; *therefore,* she was drafted to fill in there.

> We have had constant trouble with your products; *consequently,* we have contracted with another supplier.

> Your first two proposals were far too high; *however,* this one is more in line with our projections.

Determiners ● Another important class of structure words, *determiners,* is composed of several types of words. *A, an,* and *the,* the most common determiners, are often referred to as *articles.* Many pronouns and the possessive form of many nouns also function as determiners. These words are significant primarily because they signal (determine) the presence of a noun. Adjectives may come between a determiner and its noun, but the noun will follow directly. Some determiners are given below:

a	her	women's	an
his	no	the	my
the	all	your	some
this	its	many	every
Jane's	cats'	these	their

Auxiliary Verbs ● Verbs often appear with other verbs in groups of two, three, or even four. In such a verb cluster, the last verb is the main verb; the others are auxiliaries and show the tense or other characteristics of the main verb. The grammar of auxiliary verbs would take many pages to explain; however, the following discussion should help you get a feel for recognizing auxiliaries.

The three most commonly used auxiliaries can also function as main verbs with their own meanings—*to do, to have, to be:*

> to do: do, does, did
> to have: have, has, had, having
> to be: am, is, are, was, were, been, being

When used as auxiliaries, these verbs help to indicate tense or give emphasis. Another group of auxiliaries, the *modal auxiliaries,* have more complex meanings, indicating intent, possibility, obligation, and condition. The chief modals are:

can	may	shall	should	must
could	might	will	would	ought to

The following sentences show auxiliary verbs:

We *should* finish by noon.
I *might* expand our menu.
They *are* going to Boston next week.
They *have* been there three times this year.
The entire staff *has been* working overtime this month.

Notice that *been* is used once as a main verb and once as an auxiliary to the main verb *working*.

Types of Sentences

The number of structurally correct sentence patterns in English is almost infinite. Fortunately, a few short, simple patterns serve as blocks for developing the many longer, more complicated ones. If you understand these simple patterns and some of the ways in which they can be combined and expanded, you should be able to construct an almost limitless variety of sentences. The patterns are listed here for quick reference, then each is examined in detail.

Common Patterns

1. Subject–verb	Sharks eat.	**Basic pattern**
2. Subject–verb–direct object	Sharks eat fish.	**Direct object pattern**
3. Subject–verb–indirect object–direct object	We gave him trouble.	**Indirect object pattern**
4. Subject–verb–predicate noun	Mrs. Willis is superintendent.	**Predicate noun pattern**
5. Subject–verb–predicate adjective	Mrs. Willis is talented.	**Predicate adjective pattern**
6. *There*–form of *to be*–subject	There are problems.	**Expletive pattern**
7. Subject–form of *to be*–verb–(*by* _____)	Fish are eaten (by sharks).	**Passive pattern**

1. Basic Pattern ● The simplest pattern, and the underlying pattern for all others, consists of a subject and a verb. The subject is usually a noun, often with determiners, adjectives, or modifying phrases. The verb is generally a single-word action verb, often preceded by auxiliaries, with adverbs or modifying phrases. The following sentences demonstrate variations of this basic pattern:

Employees loaf
(noun)   (verb)
Three employees frequently loaf around the water cooler.
(adj.)   (adverb)   (prepositional phrase)
The   three new employees never loaf excessively.
(det.) (adjs.)   (adv.)   (adverb)

2. Direct Object Pattern ● The direct object pattern consists of a noun, an action verb, and another noun. All may have modifiers. The first noun is considered the *doer* of the action; the second is the *receiver,* or the *direct object.* Action verbs that function with direct objects are called *transitive* verbs; verbs such as those in Pattern 1 that do not need direct objects are called *intransitive.* The following examples demonstrate the direct object pattern:

> Therapists give   treatments.
> (doer)   (transitive verb) (direct object)
> Respiratory therapists often give IPPB treatments to seriously ill patients.
> (adj.)   (noun)   (adv.) (v.) (adj.) (noun)   (prepositional phrase)

3. Indirect Object Pattern ● The indirect object pattern consists of the direct object pattern with another noun, called the *indirect object,* preceding the direct object. This noun indicates to whom or occasionally for whom the action is performed. One good way to identify an indirect object is to substitute the phrase "to _____" after the direct object (as in the second sentence below). Notice the examples:

> Rodriguez sent   Jones three samples.
> (subject)   (verb) (ind.   (direct
>   obj.)   object)
> Rodriguez sent three samples to Jones.
>   (direct   (prepositional
>   object)   phrase)
> The assistant controller advanced us the necessary funds.
>   (ind.   (direct
>   obj.)   object)

4. Predicate Noun Pattern ● The predicate noun pattern differs from the first three primarily in that it contains a linking rather than an action verb. It essentially shows that the individual or group indicated by one noun can also fit the group indicated by the second noun. This enables us to classify items, and is thus a basic tool in much of our reasoning. Note the following examples:

> Lakuschievitz is   chairman.
> (subject)   (verb) (predicate noun)
> German shepherds are generally the most reliable guard dogs.
> (subject)   (verb)   (predicate noun)
> Air layering was our primary method of propagation.
> (subject)   (predicate noun)

5. Predicate Adjective Pattern ● The predicate adjective pattern contains a linking verb and a subject, but also includes an adjective showing some quality of the subject. Examples follow:

Our equipment will be outdated.
   (subj.)   (verb) (predicate adjective)
The new apprentices seem very eager.
     (subject)   (verb)   (predicate adjective)
The circuit   was complete.
   (subject) (v.) (predicate adjective)

6. Expletive Pattern ● Though not used quite as often as the other patterns in building longer sentences, the expletive pattern is extremely common. Do not overuse it, as it is usually both wordier and less direct than other patterns. Compare, for example, the first and second sentences below:

There were seventy-five employees laid off last week.
   (form of *to be*)   (subject)
Seventy-five employees were laid off last week.
      (subject) (v.)
There will be   trouble.
   (form of (subject)
   to be)
There is ample precedent for the charges.
   (form of (subject)
   to be)

7. Passive Pattern ● The passive pattern is essentially a reversal of the direct object pattern, with primary emphasis shifted from the doer to the receiver of the action. Thus the passive pattern can be useful in directing the reader's attention where you want it. But, as with the expletive pattern, it is less direct than the active patterns it substitutes for, and it is often wordier. Examples follow:

The long-range forecast will be   submitted sometime next week.
      (subject) (form of (v.)
       to be)
The statistical work was   done by Mr. DeKleva.
    (subj.) (form of (verb) (by ———)
      to be)

Expanded Sentences

Although the common patterns presented here are the basis for almost all English sentences, the overwhelming majority of sentences are composed either of combinations of two or more simple patterns or of a simple pattern expanded by the addition of various types of modifiers.

Compound Sentences ● Perhaps the simplest method of expansion through combination is to join two or more simple patterns into a compound sentence. A compound sentence consists of *two or more independent clauses.* Independent clauses are complete simple sentence patterns that could be written as independent, grammatically correct sentences. The two or more independent clauses are combined into one sentence by a coordinating conjunction (such as *and*) preceded by a comma; by a conjunctive adverb (such as *however*) preceded by a semicolon and followed by a comma; or by a semicolon alone. Joining two independent clauses by only a comma results in the error known as *comma splice.* The following examples illustrate the various means of forming compound sentences:

> The resistance at junction C was 12**, but** the resistance at junctions A and B was only 9.
> The resistance at junction C was 12**; however,** the resistance at junctions A and B was only 9.
> The resistance at junction C was 12**;** the resistance at junctions A and B was only 9.
> His white blood count was 29,000**, so** they held him for further tests.
> She was identified by three eyewitnesses**;** she had no alibi**;** her fingerprints were found at the scene.
> The Orlando market for $250 coats is very limited**; moreover,** interest in cashmere has never been high there.

Complex Sentences ● Complex sentences, like compound sentences, contain two or more clauses; however, only one of the clauses is independent. The others are dependent, or subordinate, clauses. Dependent clauses cannot stand alone because they are introduced by relative pronouns (like *who*) or by subordinating conjunctions (like *because*). They modify all or part of the independent clause and would not make grammatical sense if written alone. A dependent clause written as a separate sentence is the most common form of the so-called *sentence fragment.* In the following examples, the function of each dependent clause appears in parentheses at the end of the sentence.

> The sales representative *who will service your account* is being transferred from our Buffalo branch. (*adjective*)
> Our new assistant manager, *who has only been with our company for three months,* is very easy to work for. (*adjective*)
> *Although the Clameth Falls area is not in our district,* we will still send emergency vehicles as needed. (*adverb*)
> She was chosen *because she was the best qualified.* (*adverb*)
> You can send it to *whomever you want.* (*noun*)

Compound-Complex Sentences ● As the name implies, these sentences are both compound (having two or more independent clauses) and complex (having at least one dependent clause). Note these examples, in which the dependent clauses are italicized:

The sales representative *who will service your account* is being transferred from our Buffalo branch; however, I will be able to help you *until he arrives.*
You can send it to *whomever you want;* but don't return it to me, *because I have no use for it.*

Phrases ● The common, simple sentence pattern may also be expanded by adding various types of phrases. Even though it may be long, the resulting sentence is called a simple sentence because it contains one independent clause and no dependent clauses. Technically, many varieties of phrases exist. For instance, a noun followed by a group of modifiers is called a *noun phrase,* and a verb combined with auxiliaries and adverbs is termed a *verb phrase.* Here, though, we will only consider two types of phrases commonly used to expand sentence patterns: *prepositional phrases* and *verbal phrases.*

Prepositional phrases may function as adjectives or adverbs. They can be used in any of the roles normally served by either part of speech. Several prepositional phrases can even be used in one sentence. However, avoid stringing one such phrase after another into a "house-that-Jack-built" structure. The first example below demonstrates this ineffective stringing of prepositional phrases. The remaining examples show effective use of prepositional phrases:

The customer *in front of the counter next to the windows by the side door of the ready-to-wear department* is behaving suspiciously. (house-that-Jack-built structure)
The customer *in the yellow-striped blouse* should be watched *for possible shoplifting.*
Investment possibilities *in local condominiums* should improve *in the coming year.*

Like prepositional phrases, verbal phrases are most frequently used as adjectives or adverbs. Several specific types can be identified. *Infinitive* phrases consist of *to* plus a verb form: *to run, to lose, to read, to be.* The following examples show infinitive phrases used first as adjectives, then as adverbs, and finally in their other function as nouns:

The record *to match* was set by Unit 39 last month.
     (adjective)
The diodes *to check* are over by the telephone.
     (adjective)
We will be happy *to assist* in the project.
      (adverb)
Let's wait *to see* the completed model.
      (adverb)
To complete the job on time must be our top priority.
(noun)
I instructed them *to shut down* immediately.
      (noun)

Most other verbal phrases are marked by the *-ing,* or *participial,* form of a verb. When used as adjectives, these phrases are called *participial phrases;* used as nouns they are called *gerund phrases.* A special type called an *absolute phrase* has the participle attached to a noun and is unconnected grammatically with the rest of the sentence. It generally tells when, how, or why something happened. Below are model sentences demonstrating the use of absolute phrases, gerund phrases, and participial phrases:

> *The contracts having been signed,* we can take possession immediately.
> (absolute phrase)
> Shrinkage control must be improved, *shoplifting having increased by 33 percent.*
> (absolute phrase)
> *Improving community relations* is our top priority item.
> (gerund phrase)
> The executive committee is impatient with *your constant equivocating.*
> (gerund phrase)
> *Having increased by 33 percent,* shoplifting is now our most serious problem.
> (participial phrase)
> We found the house *burning out of control.*
> (participial phrase)

A final type of phrase to use in expanding simple sentences is the *appositive.* This is a noun cluster (a noun with modifiers) used to clarify or explain another noun. Note the examples below:

> Our proposal will be presented by our chief technician, *a skilled public speaker.*
> (appositive phrase)
> The county pollution control board, *a group of seven concerned citizens,* will hear our appeal next week. (appositive phrase)

Using Expansion Effectively ● As previously stated, good writers use the various techniques mentioned here in order to develop effective sentences. Using many unexpanded sentences, consisting of only common patterns and single-word modifiers, is rarely effective. An occasional short, unexpanded sentence can be effective, but a steady string of them will bore almost any sophisticated reader and will stamp your style as choppy or immature. Furthermore, you will be unable to show subtle relationships between ideas and communicate those ideas concisely.

However, merely making sentences long does not make them effective. Rambling or stringy sentences, consisting of one clause or phrase after another strung together with conjunctions, can be just as difficult to read as short, choppy ones. In fact, many grammatical and punctuation errors occur because a writer is trying to string too much together in one sentence.

No one length or type of sentence is automatically good or bad. Every good writer, whether an often-published professional or a specialist in another field who handles writing chores well, has an individual style, different in large part from that of others because of variations in sentence type and length. As your writing skill matures, you too will develop an individual style marked in part by the type of sentence expansions you favor. Therefore, consider the advice below as just that—advice. Practice various types of expansions. Learn to write various types of sentences. Gradually you will develop favorites.

1.   Do not overuse unexpanded, "primer" sentences. An occasional short sentence, even if only two or three words, will stand out and give good emphasis. A steady stream of such sentences will sound like an elementary-school reading text.

2.   Write fully developed expansions, but do not attempt to create extraordinarily long sentences. Gifted professionals can compose extremely long sentences, but the amateur risks grammatical errors and loss of clarity. Eighteen to twenty-two words is probably a good average, but individual sentences can vary from a few words to thirty or more.

3.   Experiment with various types of expansions. Use phrases and clauses, coordinating and subordinating conjunctions, single-word modifiers and phrase modifiers. Use each of the common patterns, avoiding overreliance on any one.

Below are some typical expansions:

1.   **a.**   He was named chairman.
  b.   Owning 40 percent of the company's stock, he was named chairman of the board.
  c.   He was named chairman on the third ballot, the first two having ended in ties.
2.   **a.**   We elected him chairman.
  b.   After two inconclusive votes, we finally elected him chairman.
  c.   We finally, after two inconclusive votes, elected him chairman of the board for the next three fiscal years.
3.   **a.**   The woofer is for low frequencies. The tweeter is for high frequencies. The mid-range speaker is for mid-range frequencies.
  b.   The woofer is for low frequencies, the tweeter for high, and the mid-range speaker for mid-ranges.
  c.   The woofer, often as much as ten inches high, is for low frequencies; the slightly smaller mid-range speaker is for middle-range sounds, with the tiny tweeter for high frequencies.
4.   **a.**   Begonias like slightly dry soil.
  b.   Begonias like slightly dry soil and can even tolerate mild drought.
  c.   Because of their rough, gnarled roots, begonias like slightly dry soil, even tolerating mild drought.

16

Avoiding Common Grammatical Errors

Sentence Fragments

Run-on Sentences and Comma Splices

Agreement: Subject and Verb

Agreement: Pronoun and Antecedent

Reference: Pronouns

Modifiers: Dangling or Misplaced

Faulty Parallelism

Pronoun Case

Overview   *Grammatical errors can cause misunderstanding, distract the reader, and give a bad impression. If you can recognize independent clauses, you should be able to avoid writing sentence fragments, run-on sentences, and comma splices. Make certain that your verbs agree with their subjects, that pronouns agree with their antecedents, and that each pronoun refers clearly to the noun it stands for. Learn to distinguish among the three pronoun cases and to make the case of each pronoun fit the function of that pronoun in the sentence.*

Most grammarians agree that the term *grammatical error* is applied too loosely to a number of expressions that are not ungrammatical but that go against the way most people think educated people should talk and write. Certainly some errors of this kind actually keep your reader from getting your message. But even when they do not cause misunderstanding, the reader will look less favorably on you and your message if you make what he or she has learned to think of as grammatical errors.

Sentence Fragments

As stated in the last chapter, *sentence* can be defined in many ways, with no commonly used definition completely satisfactory. Rather, the writer of good sentences can usually develop a feel for sentences—*sentence sense.* An important aspect of the sentence sense needed by any effective writer is the ability to recognize and avoid using sentence fragments—groups of words mistakenly treated as complete sentences. Although they begin with a capital letter and end with a period, question mark, or exclamation point, they lack at least one of the grammatical elements necessary to a proper sentence.

Perhaps the best rule of thumb to follow in checking your writing for fragments is that *a grammatical sentence must have at least one independent clause.* Nearly all fragments therefore consist of a dependent clause written alone or of a phrase or group of phrases written alone. These fragments are most often the result of the writer's chopping the dependent clause or phrase from the beginning or ending of an adjoining complete sentence. Correction, as you can note below, is usually a simple matter. The following examples show the most common types of clauses or phrases used incorrectly as fragments, followed by corrections.

1. After three months of searching for the best possible president, the board of directors finally chose Jason Wheelwright. *A thirty-year company employee, currently managing the Des Moines branch. (appositive phrase as fragment)*
After three months of searching for the best possible president, the board of directors finally chose Jason Wheelwright, *a thirty-year company employee, currently managing the Des Moines branch. (corrected)*
2. We had good reason to be disappointed. *Having worked on the project for six months. (participial phrase as fragment)*

We had good reason to be disappointed, *having worked on the project for six months.* *(corrected)*

3.   Since miniaturization has made microcircuitry possible. We should consider remodeling our system to make better use of available space. *(introductory dependent clause as fragment)*

Since miniaturization has made microcircuitry possible, we should consider remodeling our system to make better use of available space. *(corrected)*

4.   The department will have to have the specifications for the pumper redrawn and resubmitted. *With more care exercised to include all auxiliary system and options.* *(prepositional phrase as fragment)*

The department will have to have the specifications for the pumper redrawn and resubmitted *with more care exercised to include all auxiliary systems and options.* *(corrected)*

5.   All their efforts were directed toward one goal. *To attain the largest possible market share.* *(infinitive as fragment)*

All their efforts were directed toward one goal: *to attain the largest possible market share.* *(corrected)*

Run-On Sentences and Comma Splices

A second common error in sentence construction is the *run-on,* or *fused, sentence*—two or more otherwise acceptable sentences written together as one. The fragment most often occurs when a part of a sentence is chopped off with a period or other mark of end punctuation used improperly; the run-on occurs when several sentences are not separated and necessary end punctuation is omitted (Ms. Peterson could not locate the report it was buried under a pile of papers on her desk). To identify and avoid run-ons you must learn to recognize independent clauses and remember that *two independent clauses cannot be written together without proper punctuation. Several types of punctuation can be used:*

1.   Two independent clauses can be joined by a semicolon.
2.   Two independent clauses can be joined by a coordinating conjunction such as *and, but, or, nor, yet,* or *so.*
3.   Two independent clauses can be joined by a conjunctive adverb such as *however, furthermore, instead,* or *moreover.* The adverb is preceded by a semicolon and followed by a comma.
4.   Two independent clauses can be separated into two sentences with a period, question mark, or exclamation point after the first clause and a capital letter starting the second.
5.   Two independent clauses can be joined by converting one of them into a dependent clause.

Two independent clauses *cannot* be properly joined with a comma. Doing so results in the common sentence error referred to as a *comma splice,* or *comma fault.* While usually considered less serious than the run-on, the comma splice can sometimes cause confusion, and it is readily recognizable by many business and professional people as an error. Avoid the comma splice in all except extremely short, simple sentences.

The examples below demonstrate a variety of run-ons and comma splices, corrected in the ways suggested above.

1. Mr. Kozelko fell attempting to climb out over the bedrail he will have to be restrained. *(run-on)*

Mr. Kozelko fell attempting to climb out over the bedrail, he will have to be restrained. *(comma splice)*

Mr. Kozelko fell attempting to climb out over the bedrail; he will have to be restrained. *(corrected with semicolon)*

2. Our Coulter Counter should be replaced it is beyond repair. *(run-on)*

Our Coulter Counter should be replaced, it is beyond repair. *(comma splice)*

Our Coulter Counter should be replaced as it is beyond repair. *(corrected by conversion to dependent clause)*

3. Lance is an excellent product Mace and several other brands might also meet our needs. *(run-on)*

Lance is an excellent product, Mace and several other brands might also meet our needs. *(comma splice)*

Lance is an excellent product, but Mace and several other brands might also meet our needs *(corrected with coordinating conjunction)*

4. Next week we should topdress the sixteenth and eighteenth greens seventeen can wait another week. *(run-on)*

Next week we should topdress the sixteenth and eighteenth greens; however, seventeen can wait another week. *(corrected with conjunctive adverb)*

5. Joel has double-checked the program and cannot find the problem, I'll have to go over it myself. *(comma splice)*

Joel has double-checked the program and cannot find the problem. I'll have to go over it myself. *(corrected by conversion to two sentences)*

6. The IBM repairman arrived, the telephone man was here, and three salesmen were also around. *(generally accepted use of commas)*

The IBM repairman arrived; the telephone man was here; three salesmen were also around. *(safer, more formal punctuation)*

You probably noticed that several of the sample sentences could have been corrected in other ways. Once you can consistently identify and correct run-on sentences and comma splices, you can find more subtle ways of revising, reconstructing the sentences in various effective patterns.

Agreement: Subject and Verb

Inexperienced writers often make the mistake of using a plural verb with a singular subject (Politics have always interested me), or a singular verb with a plural subject (The fan and all adjoining ductwork has to be cleaned out). Obviously, the two should agree in number: singular with singular, plural with plural.

Many errors in verb agreement occur because the writer does not recognize the subject of an expanded sentence and makes the verb agree with some other word. Studying a sentence carefully to identify the basic sentence pattern will often enable you to determine the correct forms to use. The three examples below demonstrate how to correct errors by locating the basic sentence pattern.

1. The suggestion from the board, our executive committee, and the union leadership *were* accepted. *(incorrect)*
The *suggestion was* accepted. *(correct)*
Note that the basic pattern clearly requires the singular verb *was*. Modifiers used to expand the sentence have no effect on the subject and verb.
2. Our first priority in cutting expenses *are* reductions in personnel costs. *(incorrect)*
Our first *priority* in cutting expenses *is* reduction in personnel costs. *(correct)*
3. There *are* a long list of other money-saving suggestions to consider. *(incorrect)*
There *is* a long *list* of other money-saving suggestions to consider. *(correct)*

Another common mistake in verb agreement is a singular verb used with two singular subjects joined by *and.* With one primary exception mentioned below, when you see *and* joining two subjects (singular or plural), use a plural verb:

The *preparation* of a budget and constant *auditing* of each department's books *constitutes* the bulk of his job. *(incorrect)*

The primary exception to this principle is the subject that includes *each* or *every.* These two words require a singular verb regardless of the rest of the subject:

1. Every man, woman, and child *risk* drowning when swimming alone. *(incorrect)*
Every man, woman and child risks drowning when swimming alone. *(correct)*
2. Each of our junipers and balsam trees *are* on sale at half price. *(incorrect)*
Each of our junipers and balsam trees is on sale at half price. *(correct)*

One of the more confusing situations for determining verb agreement occurs when *or* or *nor* joins two or more subjects. If both subjects are singular, a singular verb is needed; if both are plural, a plural verb is needed; when one subject is singular and the other plural, the verb agrees with the nearer subject:

1. Either the yew or the pittosporum *is* going to need to be moved. *(correct)*
2. Neither the blood specimens nor the urine specimens *are* ready for analysis. *(correct)*
3. Either the disc pack or the tapes *are* available. *(correct)*
4. Neither the engineers nor the lieutenant *is* busy *(correct)*

In addition to *each* and *every,* the following indefinite pronouns require singular verbs: *somebody, nobody, everybody, someone, either, neither, anyone.* Note the following examples:

1. Each of the squad's thirteen members *is* to be given a special monthly bonus. *(correct)*
2. Anyone who applies *is* guaranteed an interview. *(correct)*
3. Neither of the drawings *needs* revision. *(correct)*

A special group of nouns, usually termed *collective nouns,* can be used with either a singular or plural verb depending on your intentions. Such nouns refer to a group or collection of individuals; a singular verb stresses the individuals, a plural verb the group. The following are a few commonly used collective nouns: *class, committee, team, family, crew, jury, staff, gang.* The examples demonstrate collective nouns used in the singular and in the plural:

1. The code *team remains* on call twenty-four hours a day. *(singular, refers to team as a unit)*
2. The code team *are* all skilled professionals with differing specialities. *(plural, refers to individuals)*
3. The jury *is* sequestered throughout the trial. *(refers to the group)*
4. The work *crew decide* each Monday which jobs they want to do for the week. *(individuals)*
5. The steering *committee meets* each Wednesday at 9:30 A.M. *(group)*

Another special case involves nouns that look like plurals but function in the singular. When they do, they require singular verbs. *Mathematics, statistics, economics, politics, checkers, measles,* and *physics* are common words that look plural but generally function as singular. The following examples demonstrate their effective use:

1. Politics, unfortunately, often *enters* into committee decisions. *(singular)*
2. Although measles *was* a serious menace to the unborn only a few years back, our inoculation program has almost eliminated it. *(singular)*

Agreement: Pronoun and Antecedent

A pronoun generally takes the place of a noun, which is referred to as the pronoun's *antecedent.* A common error made by unskilled or careless writers is failure to make a pronoun agree in number (singular or plural) with its antecedent (Each employee will have to complete their project on time). Such errors occasionally create misunderstanding or ambiguity; more often they strike careful users of English as signs of the writer's lack of concern.

Three situations require special care in pronoun-antecedent agreement. The first involves the use of the collective nouns previously mentioned. When a collective noun such as *committee* or *team* is used with a singular verb it needs a singular pronoun; when it is used with a plural verb it requires a plural pronoun. Note the examples below:

1.   The committee *is* reaching *their* own decision. *(incorrect)*
The committee *is* reaching *its* own decision. *(correct)*
2.   The team *are* going to do *its* level best. *(incorrect)*
The team *are* going to do *their* level best. *(correct)*

When *and* is used to join two or more antecedents, they are always referred to by a plural pronoun. When *or* or *nor* joins two singular antecedents, a singular pronoun is used. When *or* or *nor* joins one singular and one plural antecedent, the pronoun must agree with the closer antecedent. The following sentences illustrate these points.

1.   Air layering and simple layering are quick and easy; use *them* whenever possible. *(correct)*
2.   Either the foreman or the shift supervisor will be at *his* station. *(correct)*
3.   Either the foreman or the hourly employees will be at their stations. *(correct)*

The third situation requiring careful matching of a pronoun with its antecedent occurs when words such as *person, somebody, neither, another, anyone, someone, one, no one,* or *nobody* are used as antecedents. The following examples demonstrate the problems involved.

1.   When one is careful, *he* will make few mistakes. *(correct)*
2.   Just because someone makes a minor mistake, *he* should not be fired. *(correct)*

Notice that, although each of the two examples just given could as easily refer to a female as to a male subject, the masculine pronoun form has been used. Traditionally the masculine forms *(he, his, him)* have been required in such cases. Recent practice, however, is to avoid such usage. Three possible methods can be used: In informal writing you can safely use the plural form *(they, their, them);* or you can even use *he/she, his/her,* or *him/her;* but in formal writing these usages are still considered inappropriate, so you should recast the entire sentence. Note the examples below:

1.   When one is careful, *they* will not make mistakes. *(informal)*
Careful employees will not make mistakes. *(formal)*
2.   Just because someone makes a simple mistake, *he/she* should not be fired. *(informal)*
Making a simple mistake is not sufficient cause for dismissal. *(formal)*

Reference: Pronouns

Many sentence problems occur when a pronoun does not have one readily recognizable antecedent (the noun replaced by a pronoun). A common variety of this problem is ambiguous reference, where the pronoun may refer to two or more equally possible antecedents. The examples below show how to correct ambiguous pronoun reference:

1. My brother's attorney said that he would soon be contacted by the court. *(ambiguous—who will be contacted?)*
My brother was told by his attorney to expect to be contacted by the court soon.
My brother's attorney said, "I'll soon be contacted by the court."
2. My resume's structure is so effective that anyone should be impressed by it. *(ambiguous—the resume? or its structure?)*
The structure of my resume is effective enough to impress anyone.
With such an effective structure, my resume should impress anyone.

An equally bothersome reference problem is the remote or vague reference, which occurs when there is no readily obvious antecedent. To avoid this problem, place the pronoun as close as possible to its antecedent—not necessarily in the same sentence, but close enough to be clearly noted. The following examples show some typical vague references and possible corrections.

1. The collator was repaired by the George Stuart Company serviceman, who came within two hours of our call. *It* needed only a minor adjustment. *(remote)*
The collator was given the needed minor adjustment by the George Stuart Company repairman, who came within two hours of our call. *(correct)*
2. When the machine's defective circuitry was repaired, *it* was soon put back into operation. *(remote)*
The machine was put back into operation soon after its defective circuitry was repaired. *(correct)*
3. After Bernie's new job opens up *he* will have to move. *(remote)*
Bernie will have to move after his new job opens up. *(correct)*

Another type of vague reference is what is sometimes called *broad reference*. A pronoun such as *it, this, that, which,* or *such* is used carelessly so that the reader cannot at first be certain whether the pronoun refers to a specific word or phrase or to an entire idea. These pronouns may be used correctly in this broad sense, but the inexperienced writer is safer to use them only to refer to obvious antecedents. Note the vague, broad reference in the following sentences and the suggested revisions.

1. Staining a slide properly is basic to all micro-lab work and should be learned and practiced regularly. *It* is relatively simple but extremely important. *(broad)*
Staining a slide properly is basic to all micro-lab work and should be learned and practiced regularly. The process is relatively simple but extremely important. *(correct)*
2. The arson investigation was completed by Inspector McLuan, but Chief Kaiser worked with the grand jury and the states attorney, *which* is commonly done. *(broad)*
Following common practice, Inspector McLuan conducted the arson investigation while Chief Kaiser worked with the grand jury and the states attorney. *(correct)*

A similar type of vague reference results from careless use of the pronouns *it, you,* or *they* when no clear-cut antecedent is given. Use *you* only when you are speaking directly to the reader. Use *it* to refer to a definite antecedent. The only exception is the generally accepted expletive pattern:

1. It is cold in here. *(acceptable—expletive pattern)*
2. It is time to go. *(acceptable—expletive pattern)*
3. In our policy manual *it* says that we are entitled to negotiate individual salaries. *(vague)*
Our policy manual gives us the right to negotiate our own salaries. *(correct)*
4. When we planned the Winter Oaks development, *they* hadn't even built the Crosstown Expressway. *(vague)*
When we planned the Winter Oaks development, the Crosstown Expressway hadn't even been built. *(correct)*
5. The MG transmission is both very strong and very complicated to work on. A car builder should leave it alone. *You* can get into needless trouble if *you* tamper with it. *(vague)*
It can cause needless trouble if tampered with. *(correct)*

Modifiers: Dangling or Misplaced

The misplaced or dangling modifier is one of the most common sentence problems. At times it will go unnoticed by any but the most perceptive reader; other times it will give a humorous or absurd misreading; and occasionally it will create a genuine ambiguity. Careful writers cannot afford unintentional ambiguity, nor can they afford to weaken their writing by allowing an absurd misreading of it. Most dangling modifiers result from introductory phrases (usually participial phrases) which modify something other than the subject of the sentence. The reader will assume unconsciously that an introductory phrase refers to the subject of the sentence and can thus be either confused or amused. The examples below show dangling modifiers and their correction:

1. Entering the rear door stealthily, the suspects were caught unawares. *(ambiguous: was it the suspects or the police who entered?)*
Entering the rear door stealthily, the police caught the suspects unawares.
Entering the rear door stealthily, the suspects walked right into a trap. *(correct)*
2. Minding my own business, the machine stopped suddenly. *(absurd misreading)*
I was minding my own business when I heard the machine stop. *(correct)*
3. In order to meet our schedule, each delivery must be made in less than fifteen minutes. *(dangling)*
In order to meet our schedule, we must make each delivery in less than fifteen minutes. *(correct)*
4. While greeting prospective clients, my cigarette burned a hole in my desk. *(dangling)*
While I was greeting prospective clients, my cigarette burned a hole in my desk. *(correct)*
5. Before leaving for the inspection trip, our itineraries were carefully planned. *(dangling)*
Before we left for the inspection trip, we planned our itineraries carefully. *(correct)*

Instead of dangling alone at the beginning of a sentence, a misplaced modifier is misplaced within a sentence and either seems to modify the wrong word or may modify either of several words. Word order greatly influences the meaning of sentences; the same modifier moved around in a sentence can be made to modify several different words, each time changing the overall meaning of the sentence. Notice in the following series of sentences how moving the adverb *only* changes the meaning:

1. The dentist only asked his assistant to hand him some di-cal.
2. The dentist asked only his assistant to hand him some di-cal.
3. The dentist asked his only assistant to hand him some di-cal.
4. The dentist asked his assistant only to hand him some di-cal.
5. The dentist asked his assistant to hand only him some di-cal.
6. The dentist asked his assistant to hand him only some di-cal.

Position your modifiers carefully, and reread your sentences to determine that they have exactly the meaning you want. The following examples demonstrate ambiguity or absurdity resulting from misplaced modifiers. Notice how each is corrected.

1. The entire crew has followed the instructions *precisely* given in the operational manual. *(ambiguous)*
The entire crew has *precisely* followed the instructions given in the operational manual. *(clear)*
The entire crew has followed the instructions given *precisely* in the operational manual. *(clear)*
2. They stopped pouring concrete after they discovered a sinkhole under the foundation *that had* apparently been there for years. *(confusing)*
They stopped pouring concrete for the foundation after discovering a sinkhole *that had* apparently been there for years. *(clear)*
3. The patrolman who broke his arm *yesterday* returned to work. *(confusing)*
The patrolman who *yesterday* broke his arm returned to work. *(clear)*
The patrolman who broke his arm returned to work *yesterday. (clear)*

Faulty Parallelism

Parallel structure is the placement of two or more words, phrases, or clauses in series, using commas or conjunctions. The following sentences demonstrate parallel structure.

1. He searched for the missing blueprint in the desk, in the file drawer, and even in the wastebasket. *(parallel prepositional phrases)*
2. She ran errands till seven, shopped for groceries till eight, ate a late dinner till nine, then collapsed into bed at ten. *(parallel verb phrases)*
3. Her recreational activities included grooming her horse, waterskiing, cooking gourmet meals, and giving karate lessons. *(parallel gerund phrases)*

4.   We wanted Mr. Desmond's last day on the job to be both a reasonably typical workday and a memorable time. *(parallel noun phrases)*

Parallel structure is an important element of anyone's writing. Such structures may be short and simple or long and complex. Often you can use parallelism to coordinate related material or to show contrasts. In any case, you must be certain to use it properly. If you present two or more items as parallel you must express them in similar grammatical structure: use a participial phrase only with other participial phrases, infinitives only with infinitives, and so on. The following examples show faulty parallelism followed by corrections.

1.   New employees must be shown how to fill out timecards and sign-in procedures. *(faulty)*
New employees must be shown how to fill out timecards and sign in. *(correct)*
New employees must be shown proper procedures for filling out timecards and for signing in. *(correct)*
2.   Ms. Fletcher always performed her duties rapidly, efficiently, and with a smile. *(faulty)*
Ms. Fletcher always performed her duties rapidly, efficiently, and smilingly. *(correct)*
3.   Seagraves is not only well liked as a supervisor but also because he is a nice guy. *(faulty)*
Seagraves is well liked not only as a supervisor but also as a friend. *(correct)*
4.   Colcer asked her staff to treat each client as their only client and that they should give each one as much personal attention as possible. *(faulty)*
Colcer asked her staff to treat each client as their only client and to give each one as much personal attention as possible. *(correct)*

Pronoun Case

As mentioned earlier, personal pronouns have a characteristic termed *case:* nominative case, objective case, or possessive case. The appropriate case is determined by the pronoun's function in the sentence. The nominative case is required for subjects and predicate nominatives; the objective case is used for objects; the possessive case is used to show possession and functions almost as an adjective. The sentences below illustrate basic case forms.

1.   *They* caused excessive flooding of the drainfield. *(nominative case as subject)*
2.   *Their* system gives *them* trouble. *(possessive case/objective case as indirect object)*
3.   The first amniotic tap was conducted in *her* eighth month. *(possessive)*
4.   This is *he*. *(nominative case as predicate noun)*
5.   *We* sent *him* three reminders of *his* overdue bill. *(nominative/objective/possessive)*

	Nominative	Objective	Possessive
First person	I, we	me, us	my, mine, our, ours
Second person	you	you	your, yours
Third person	he, she, it, they	him, her, it, them	his, her, hers, its, their, theirs

The example sentences below illustrate the more common errors in pronoun case and their correction:

1. For a long while it seemed as if *him* and *me* were going to be blamed. *(subject needs nominative case)*
For a long while it seemed as if *he* and *I* were going to be blamed. *(correct)*
2. Just between *you* and *I*. . . . *(object of preposition needs objective case)*
Just between *you* and *me*. . . . *(correct)*
3. Not many of *we* trainees are left. *(object requires objective case)*
Not many of *us* trainees are left. *(correct)*
4. The executive board has selected three candidates—Stubbs, Malone, and *I*. *(objective case needed because appositive must agree with word it refers to)*
The executive board has selected three candidates—Stubbs, Malone, and *me*. *(correct)*
5. No other therapist in town is as skillful as *him*. *(nominative required for subject of understood verb* is skillful*)*
No other therapist in town is as skillful as *he*. *(correct)*
6. Management is concerned about *him* working two jobs. *(possessive required to modify the gerund* working*)*
Management is concerned about *his* working two jobs. *(correct)*

Interrogative and relative pronouns also have case and must therefore be chosen carefully. The principles for using them are the same as those for using personal pronouns. The only really troublesome interrogative or relative pronoun choice is between *who* and *whom*. Current practice is to ignore *whom* except in formal speech or writing. Keeping in mind, though, that technical readers are generally conservative in matters of grammar, usage, and form, your safest course is to retain the distinction: *who* for nominative positions, *whom* for objective positions. The examples below illustrate proper use of *who* and *whom*:

1. *Who* is there? *(used as subject)*
2. *Who* did you say is calling? *(subject of* is calling*)*
3. Victor is the one *whom* we should elect. *(object of* should elect*)*
4. He is one in *whom* we can place our trust. *(object of* in*)*
5. Sales should increase no matter *who* is chosen. *(subject of* is chosen*)*

The eight problems in sentence construction discussed in this chapter do not exhaust the errors recognized by some very careful readers, but they are the most common and most troublesome. If you master them, you can be reasonably confident that your sentences are conveying the right impression as well as the right message to your readers.

17

Punctuation, Spelling, and Mechanics

Spelling

  Capitalization

  Apostrophes

  Abbreviations

Using Numbers

Manuscript Form

Overview   Most readers react negatively to poor or idiosyncratic spelling, punctuation, and mechanics. Punctuation shows how your sentences should be read. Every sentence requires appropriate end punctuation. Commas, which indicate a brief pause, are the most frequently used punctuation. Know when to use them, when not to use them, and when to use stronger punctuation such as semicolons, colons, dashes, and parentheses. Special care is required in punctuating quotations. If you have trouble spelling correctly, use the dictionary. Follow convention in using capital letters, apostrophes, and numerals. Follow standard manuscript form unless your organization specifies differently.

As a guide to how sentences should be read, punctuation can never imitate all the subtle tones of voice we use in conversation, but it can help. And in some cases punctuation allows us to be more exact and concise in writing than in speaking: consider how much more easily we can indicate a quotation in writing then when we have to introduce it in speech.

This chapter examines three categories of punctuation: the marks that end sentences, those that indicate various kinds of breaks within sentences, and those that show when we are quoting, emphasizing, or using words in a special way.

End Punctuation

All sentences require appropriate end punctuation. Three marks are commonly used: the period (.) ends statements or instructions; the question mark (?) ends questions; the exclamation mark (!) ends emphatic or exclamatory sentences.

Periods are by far the most common end stops. The sample sentences below illustrate common types of sentences ended with periods.

1. *Direct statement:* The bonds we wanted are now 8¾.
2. *Indirect question:* Mr. Robinson is going to ask why we did not buy them. (Notice that the sentence is a statement about Mr. Robinson's question.)

3. *Instruction:* Next, verify all departmental expense sheets.
4. *Courtesy question:* Would you please stop by my office. (Notice that although this is phrased as a question it really functions as a request or instruction.)

Periods are also required after most abbreviations:

Ms. Mr. Mrs. (not Miss) Dec. Mon. ob. peds. a.m. i.e.

Acronyms (abbreviations pronounced as words) and abbreviations for companies and governmental agencies are not punctuated:

WAVE   CBS   CIA

Abbreviations are discussed in more detail in a later section.

Periods are also used as decimal points preceding a decimal fraction, between a whole number and decimals, and between dollars and cents:

$23.95   .95   91.6%   .238   3.1416

Question marks are used as end stops for sentences that ask direct questions.

1. Why is the FM band so much clearer?
2. You actually want to apply for that job? (Notice that here the question mark shows the reader how to read the sentence; a period after the same group of words would have given a totally different reading.)
3. I want to, but how can I? (The question mark is needed because the question ends the sentence.)

Question marks can sometimes be used within sentences:

"Will we finish on time?" the board asked.
He called in three backup units (who knows why?) when none were needed.

Exclamation marks show extreme emphasis and are rarely needed in job-related writing. Use them only with interjections such as Oh! Help! Fire! or with statements or commands that must convey strong emotion. Ordinarily you can achieve sufficient emphasis through effective choice of words.

Commas

Commas are the most used—and most misused—marks of punctuation. You should learn not only when commas are needed, but also when a stronger mark of punctuation is called for and when no mark at all is required. Commas are used within sentences, marking separations between grammatical units. They indicate a slight pause to the reader, much like a brief pause in speech or a short rest in music. When a more definite, longer pause is indicated, a stronger mark is needed; when no pause is indicated, no mark is called for. The first six sections below explain and illustrate the principal uses of commas; the last one shows some situations in which commas should not be used.

Commas Between Certain Clauses

In compound sentences with clauses joined by a coordinating conjunction, a comma should precede the conjunction unless the clauses are short and closely related:

1. The excursion trade is booming this year, and we would be fools not to take advantage of it.
2. The entire area is thriving and we're right in the middle. *(comma may be omitted with short clauses)*
3. We realize that space in the detention center is limited, but we feel that you could at least use bunk beds instead of putting mattresses on the floor.
4. Conditions do seem intolerable there, yet we have no choice but to tolerate them.
5. These plants will outgrow the terrarium in a few months, so will you just leave them in the pots when you plant them?

A subordinate clause placed at the beginning of a sentence is usually followed by a comma. Only if the clause is extremely short and closely related to the main clause can the comma safely be omitted:

1. After spending three months preparing our proposal, we were flabbergasted when it was turned down within a week.
2. When we installed new photocopiers in every secretarial cluster, we had no idea they would be used so much.
3. After completing my speech I collapsed. *(comma may be omitted)*

Subordinate clauses that follow the main clause are seldom separated out by commas. Put a comma before such a clause only when it is very loosely related to the main clause and a distinct pause in reading is called for:

1. We had no idea the new photocopiers would be used so much when we installed them in every secretarial cluster.
2. The superintendent has decided on using the CPM in developing next year's budget, although he has never even tried it on a small job.

Long phrases coming at the beginning or end of a sentence are punctuated much the same as are subordinate clauses.

1. Paying close attention to the dog's reaction, he gave a gentle tug on the training collar.

2. To make certain that the dog understood his pleasure with the behavior, he gave her a milk biscuit and praised her lavishly.
3. Being too heavy to carry, the console had to be repaired in the customer's home. *(Notice the misreading possible if this comma is omitted.)*
4. Having tried everything he quit in disgust. *(comma may be omitted)*
5. He carefully pruned the upper branches to stimulate the lower ones. *(no comma)*
6. Dennison approached each prospective customer gingerly, watching for a reaction. *(the comma prevents ambiguity)*

Commas to Separate Items in a Series

Commas are used to separate the items—words, phrases, or clauses—in a series.

1. Crandall, Buchan, Reisinger, and Pauley will be transferred to the third shift next week.
2. The problem could be either a lack of refrigerant, a malfunctioning timer, or a broken shut-off valve.
3. Preliminary indoctrination should include instruction in how to clock in and out, how to complete the hourly log, how to record all messages received, and how to submit required vouchers and chits.

Notice that a comma precedes the conjunction (usually *and* or *or*) before the final item. Some writers prefer to omit this comma, but the safest practice is to use it, since sometimes its omission can cause misunderstanding. Notice the possible confusion if the final comma below were omitted:

4. Make up and stencil drawers for each of the following items: washers, mollies, wing nuts, nuts, and bolts. *(Without the comma the reader would be uncertain whether to make up four drawers or five.)*

Do not use commas between items in a series when all are joined by conjunctions:

5. You could plant a bottlebrush tree or a punk tree or even an Australian pine there. *(no comma)*

A type of series that is especially bothersome to punctuate is the series of adjectives modifying the same noun. Commas are used to separate the items in such a series when they are coordinate; that is, when *and* could be used to join them or when their order could be reversed. Commas are not used when the adjectives are arranged so that one adjective modifies the rest of the expression, so that *and* could not be used nor the order reversed.

1. She was promoted into a mentally challenging, physically tiring, and psychologically demanding position.
2. His initial judgment was to wait a few days before operating on the anemic, obese, and obviously very nervous patient.

3. We replaced our aging, outmoded adding machine.
4. We replaced all of our old manual typewriters. *(no comma;* old *modifies* manual typewriters*)*

A comma is never used after the last adjective.

Commas to Set Off Parenthetical Elements

Commas are generally used to set off words, clauses, or phrases that interrupt the basic sentence pattern. When the interruption is very slight, the commas are sometimes omitted. However, the safest practice in job-related writing is to use the commas. If you are in doubt, you can read the sentence aloud; if pauses are called for around the expression, use commas.

1. We wanted, nevertheless, to make her as comfortable as possible.
2. Our Apopka plant, of course, has a much larger selection.
3. On the other hand, the MA-2 has been very useful for several years.
4. Our investigation, consequently, had to be postponed for a month.
5. Yes, we understand completely.
6. Denise, you will have to work overtime every night until you are caught up.
7. We are certain, colleagues, that our fall junior line will be our most successful ever.

Commas to Set Off Nonrestrictive Modifiers

Nonrestrictive modifiers are phrases or subordinate clauses that give extra information rather than limiting, or restricting, the expression they modify. Nonrestrictive elements could be omitted without significantly changing the sentence's meaning. The sentences below contain nonrestrictive modifiers.

1. Our ward clerk, *who has only been with us for three months,* saves us all a great deal of work.
2. The four-bed, two-and-one-half-bath Spanish on Tarrytown Road, *which has been vacant for six months,* is too garish to sell.
3. Great Southern Inns, *selling for 16¾,* is a great buy today.

Notice that the nonrestrictive elements set off by the commas could be removed without changing the essential meaning of the sentences. Restrictive elements, on the other hand, restrict, or limit, the meaning of the expressions they modify so that to omit them would either change or confuse the meaning of the sentence. Restrictive elements are not set off with commas. The next two sentences show restrictive modifiers:

4. Employees *who have been with us less than three months* are ineligible for personal leave.
5. The work *that we do not get done today* can be finished on Monday.

Determining whether a modifier is restrictive is often difficult. Study the sentence both with and without the expression. If the meaning is substantially different, the expression is restrictive, and no commas should be used. You can also use the reading aloud test: if your voice drops or pauses, a comma is needed.

Commas to Set Off Appositives

Appositives are nouns or noun clusters that restate and supplement a preceding noun. They are set apart from the rest of the sentence by commas:

1. Marcia Margeson, *Pantheox general manager,* will take three months' leave to run for city council.
2. Thurman Soles, *patent holder on our new dectronic leakage detector,* will be in tomorrow to show us several more new inventions.
3. When I finish my degree in horticulture, I will attend a special PGA school and then go to work at Rio Pinar, *host of the annual Citrus Open.*

Commas to Follow Convention

Many of the commas necessary in job-related writing are used because conventional practice dictates so. Commas are conventionally required, for instance, to separate the day of the month from the year:

April 16, 1968   May 16, 1972

When only a month and year are given, the comma is optional:

December, 1980   December 1980

The military form, 27 January 1979, requires no comma.

Commas are also conventionally required to separate numbers into thousands, millions, and so on:

3,816   4,861,217

Commas are conventionally used to separate items in addresses:

Riverside, California
37 West Church Street, Detroit, Michigan
4579 Kirckmon Rd., Orlando, Florida 32811

Notice that no comma is used between the state and zip code.

In sentences, the final item of an address or a date is also followed by a comma:

4415 Semoran Blvd., Orlando, is our new mailing address.
July 3, 1981, is the big day.

Commas are conventionally used to separate degrees and titles from proper names:

Rosalind D. Nelson, Ph.D.
Robert Pew, Ed.D., C.D.P.
Cheryl Johnson, C.P.S.
Wendell P. Daugherty, Jr.

Commas are conventionally used to introduce direct quotations and to separate interrupting parts of the sentence from the quotation:

1. Dr. Stahr asked, "When is her SMA-360 scheduled?"
2. "Perhaps your people could arrange a time trade," Chief Shrivers suggested.
3. "I'm not sure," I answered, "that I really want to go."

Misuse of Commas

Many inexperienced writers use far too many commas, seeming to believe that more punctuation is better than less. This is not a valid assumption. As a rule of thumb, too many commas are worse than not enough. Use a comma only when one of the guidelines cited above indicates so, or when you are positive that a comma is necessary for the sentence to read effectively.

Unnecessary commas could, of course, be placed just about anywhere, and it is impossible to show all such possible errors here. Following are some of the more common ones:

1. Our difficulties in getting the Q-Vac to fit onto a skid, must be corrected within ten days. *(incorrect—the subject is erroneously separated from its verb)*
2. Whenever we arrive more than one minute early, Chief Kaiser always jokes, that we must be going to a fire. *(incorrect—indirect quotations require no special punctuation)*
3. Conducting seminars for new employees, and giving refresher courses occupied most of his time. *(incorrect—use a comma before* and *only when joining entire clauses or when ending a series)*
4. The access roads that were completed prior to the building of Rt. 427, already need repair. *(incorrect—a restrictive clause uses no commas; a nonrestrictive clause needs two; one comma in this position is therefore never correct)*
5. Private offices, a shower room, and a lounge, are all included in the proposed new wing. *(incorrect—do not separate entire series from the rest of the sentence)*

Semicolons

The semicolon is intermediate between a comma and a period, indicating a sharper break or pause than a comma but not the complete stop indicated by a period. Semicolons have two primary uses: to connect independent clauses when no conjunction is used (a comma would be insufficient), and to replace commas in certain circumstances. The examples below show semicolons used correctly.

1. Ms. Newcombe is the most efficient underwriter we have; she also shows remarkably good judgment for one so inexperienced in writing homeowners policies. *(correct—two independent clauses)*
2. Our copier should be ready by the first of next week; however, they will not give us a loaner. *(correct—independent clauses, coordinating adverb)*
3. The exercise regimen Neumeyer gave me takes only fifteen minutes a day, but he guarantees success. *(correct—comma suffices when coordinate conjunction is used)*
4. Richard Zebby, Rob Morgenthau, and Vic Picitello are more experienced; but Richard Jones, the new man, already outsells them. *(correct—the semicolon is substituted for a comma here because of the other commas in the sentence)*
5. Las Cruces, New Mexico; Amarillo, Texas; and Golden, Colorado, are all good prospective sites. *(correct—semicolons are used when the items in a series contain other commas)*
6. The four employees honored at the banquet were Karen Steele, an accountant; Louise Miles, an expediter; Jay Feltman, a sales representative; and Terry Tyner, chief of maintenance. *(correct)*

Colons

Colons serve as markers to introduce something to follow. They should not be confused with semicolons, which separate coordinate expressions. Colons are most often used to introduce a list, an explanatory or illustrative statement, or a formal quotation. This sentence, as well as the examples below, illustrates the effective use of colons as introductory markers:

1. The ideal tree for this location should have three characteristics: strong salt tolerance; thick, heavy leaf structure; and a strong, shallow root structure. *(correct)*
2. Please send the following items on Monday's shipment: a set of HB60s, a set of HF60s, and two sets of HG70s. *(correct)*
3. As project coordinator she always followed the same policy: do what is right today; justify it later. *(correct)*

Note that in the three sentences above the colons follow grammatically complete sentences. Many inexperienced writers improperly use colons to break up a sentence.

4. The best known suppliers are: Devex Electric, Electric Industries, and Tompkins Bros. Electric. *(incorrect—no mark needed)*

In the following example, the colon is used to introduce a formal quotation:

5. His favorite statement was a quote from Churchill: "I've often had to eat my own words, and on the whole I've found them rather good eating." *(correct)*

Colons are also occasionally used between two independent clauses when the second explains, illustrates, or reiterates the first.

>   **6.**   The machine has always given us trouble: rectifier, amplifier, everything in it has malfunctioned. *(correct)*
>   **7.**   Harper will make an ideal undercover narcotics agent: he looks like an addict. *(correct)*

Dashes and Parentheses

The dash is a strong, dramatic mark that has several valid uses. Unfortunately, it is often misused and overused. In job-related writing, the dash should be used sparingly—never thrown around purely for emphasis. The two main uses of the dash are to mark a sudden change of thought or interruption in midsentence and to set off parenthetical elements when a stronger mark than the comma is needed. It is also used around parenthetical elements that contain internal commas. The examples below demonstrate effective use of dashes.

>   **1.**   Our best guess—and it is certainly just a guess—is that the final figure could reach over 5,000,000. *(correct)*
>   **2.**   We have assigned a team of our best illustrators—Fabula, Neilson, and Sternbery—to develop your project. *(correct)*
>   **3.**   Golf, tennis, swimming, boating, or hiking—Wild Deer Run is the community with everything. *(correct)*

Like commas and dashes, parentheses are used to set off parenthetical material within a sentence. No foolproof guidelines can be given to show when to use which mark; however, the following practice is conventional: commas are the most commonly used, especially for material closely related to the rest of the sentence; dashes are used for longer elements and for those with internal commas; parentheses are used for extremely long elements that are explanatory or supplementary. Parentheses enclose material that could easily be used in footnotes, material that is not essential to the meaning of the sentence. The examples below use parentheses effectively:

>   **1.**   The complete assembly cost only three hours of my labor and $29.83 for parts (see attached parts list).
>   **2.**   If you are willing to water, fertilize, and mow twice as often, Bermuda should meet your needs (Tifton 417 is a good new strain).

Punctuating Quotations

Three punctuation marks—quotation marks, brackets, and elipses—function in a different manner from most other punctuation marks. Rather than separating sentence elements and offering a guide to reading a sentence, they show when you are using someone else's expression or when you are using an expression in a special manner. None of the three is used as commonly as are most other marks, but all are especially important in presenting research data, both in college papers and in on-the-job reports.

Quotation Marks

The most important use of quotation marks is to enclose direct quotations, the exact words written or spoken by someone else. Do not use them for indirect quotations—the gist but not the wording of the other person's communication. The following demonstrate the proper use of quotation marks to enclose direct quotations:

> 1.   According to Bumby and Stimson, a job cost summary serves to "consolidate, summarize, and analyze all phase and subcontract costs."
> 2.   "Medical expenses," according to the Chamber of Commerce survey, "can include dental fees, doctor fees, medicine and drugs, nursing care, examinations and tests, medical appliances, and insurance."

Long quotations of four or more lines require special handling: single-space and indent them. Do not use quotation marks.

> Neuner and Keeling say this about organizational manuals:
> 3.   The organization manual explains the organization, the duties, and the responsibilities of the various departments and their respective divisions. Each section of the organization manual might consist of a portrayal of the duties and responsibilities of the office or department treated in that section, the organization chart, and a listing of the administrative officers of the office or department.

Quotation marks are also used to indicate titles of articles, chapters, poems, musical compositions, or paintings. (Do not put the title of a book or journal in quotation marks, however.)

> 4.   There was a very useful article entitled "Theftproof Your CB."
> 5.   The next chapter, "Cutting the Key," shows clearly how to make your own key from a blank.

Quotation marks may also be used to indicate that a word is being referred to as a word, not as a meaning; but contemporary usage more often puts such words in italics:

> 6.   She often confused "affect" and "effect."
> 7.   He is one of those perfectionists who distinguish between *oral* and *verbal*.

Some writers also use quotation marks to indicate a slang word that does not fit the tone of the sentence. A better practice is to find an appropriate word so that the quotation marks are unnecessary.

Punctuation at the End of a Quotation

Commas and periods are always placed inside closing quotation marks:

1.   "No," she replied, "he isn't ready yet."

Semicolons and colons are placed outside of closing quotation marks:

2.   In his first report he stated emphatically, "We will finish by July 30"; by the second report he had hedged a bit.

Question marks, dashes, and exclamation marks go inside of closing quotes when they refer to the quoted material, outside of them when they refer to the sentence but not the quotation:

3.   Why did they submit a formal proposal with the ridiculous title "An Immodest Proposal"?
4.   The civil service inspector asked me the typical question, "Why do you want to work for this department?"
5.   Why didn't he ask, "Why do you want a weekly paycheck?"

Note that only the inside mark is used when both the quotation and the total sentence are questions.

Ellipses

An ellipsis consists of three spaced periods. It is used to indicate material omitted within a quotation. A full line of periods may be used to show that a full paragraph or more has been omitted. Ellipses can be used at the beginning, middle, or end of a quotation. When ellipses are combined with other marks of punctuation, the other punctuation comes first.

1.   Herbert J. Gans in *New Generation* has this to say about young workers:
". . . many have embraced the expectation, common in the middle class, that the job itself should provide some satisfaction. . . ."
2.   Gans goes on to point out, "They are not saving up to buy a garage but are going to night school to become eligible for a white collar job, . . . they can express their discontent or quit when the work becomes intolerable."

Brackets

Brackets function in direct quotations to set off comments, explanations, or any other inserted material. They must not be confused with parentheses.

1.   Christman defines general aviation as "all flying done other than by scheduled carriers [airlines] and defense agencies [military]."
2.   He goes on to point out, "Private planes [general aviation] have access to over

12,000 airports in the United States alone, compared with 500 used by the airlines [actually twenty airports handle 80 percent of all our air traffic]."

Italics or Underlining

Handwritten or typed materials use underlining to indicate italic print. Use it to indicate the title of a book, periodical, play, or motion picture; to indicate a foreign word that is not yet a standard part of English; to indicate scientific names of plants, animals, diseases, and the like; to show that you are referring to a word as a word, not to what the word means. You can also use it to draw attention to a word or expression that you want to strongly emphasize, but do so sparingly. Choosing emphatic words is far better than underlining indiscriminately.

1. *Dracaena marginata* is another plant that can be easily mound layered.
2. Consolidated pulled off a *coup d'état* in gaining control of American.
3. She daily read the *Washington Post* and the *St. Louis Post Dispatch.*
4. He kept a copy of the *American Heritage Dictionary* close at hand.
5. *Guard* and *gauge* are commonly misspelled.

Spelling

Many writers mistakenly assume that spelling is a trivial part of effective writing, often using this as an excuse for poor spelling or for laziness. Spelling is important. The readers of your reports, letters, memorandums, and proposals will certainly think so.

Unfortunately, there is no magic shortcut to effective spelling. Some writers who are otherwise very skillful struggle all their careers with poor spelling. If you are a poor speller and cannot seem to improve much, allot extra time for important writing projects so that you can look up spellings in a dictionary. If you frequently use many of the same words that give you trouble, make a list for ready reference. One way or another, make certain that spelling errors do not destroy the effectiveness of your writing. The suggestions below might help you with some of the more common spelling problems; however, for many words, only memorization or reference word lists will help.

1. *ie* or *ei:* Place *i* before *e* except after *c* or when sounded like *a* as in *neighbor* and *weigh.* The rule is often helpful; however, note the following common exceptions:

height   neither   leisure   foreign   seize   weird   species

2. Final *e:* Drop a silent final *e* before suffixes beginning with a vowel; retain it before suffixes beginning with a consonant:

Write—writing	hope—hopeful	guide—guidance
love—lovely	nine—nineteen	sincere—sincerely

Common exceptions include words like *courageous* and *noticeable* in which final *e* is preceded by *g* or *c*. These retain the *e* before endings beginning in *a* or *o*. Here are some other exceptions:

dyeing   ninth   truly   awful   wholly

3. Changing *y* to *i:* When adding a suffix, change a final *y* to *i* except before a suffix beginning with *i:*

fly—flying   rely—reliance   forty—fortieth

4. Hyphenation: Hyphens are frequently used to join two words used as one word. No hard rule exists; but generally hyphens are used for newly formed compounds, while well-established compounds are written as one word. When in doubt, consult a current dictionary. Compound adjectives (two or more words used together to modify a noun) are regularly hyphenated:

matter-of-fact   far-sighted   old-fashioned   well-planned

Prefixes are generally written as part of the root words; however, those that are accented or that retain individual meaning are hyphenated:

vice-chairperson   ex-president   pro-merger

5. Commonly misspelled words: Following are some words misspelled frequently:

abundant	academic	across	additionally
adequately	adsorption	aggravate	alleviate
all right	altogether	amateur	analysis
apparent	arctic	argument	article
athletic	becoming	benefited	calendar
category	cemetery	column	committee
competition	condemn	conscientious	conscious
continuous	definitely	description	desirable
desperate	dining	disappoint	disastrous
dissatisfied	eighth	embarrass	environment
etc.	existence	exorbitant	familiar
feasible	February	forth	gauge
government	grammar	guard	harass
hurriedly	hypocrisy	imitation	incredibly
independent	infinite	intelligence	irrelevant
knowledge	laboratory	leisure	license
loneliness	magazine	maintenance	marriage
naphtha	necessary	nucleus	occasionally
occurred	omitted	parallel	permissible
personnel	possess	preceding	predominant
prejudice	prevalent	privilege	procedure

professor	prominent	psychology	pursue
recommend	repetition	ridiculous	separate
sergeant	shining	strenuous	studying
succeed	susceptible	temperament	thorough
undoubtedly	villain		

Capitalization

Capitalization is primarily a matter of convention: Once writers could capitalize any word they considered important. Now, however, we capitalize only the first word of a sentence, the word *I*, proper nouns, and important words in titles. Inappropriate capitalization will seldom obscure meaning, but it will be considered by many readers a sign of carelessness.

Capitalize the first word of any sentence, and always capitalize *I:*

1. He advised me on methods I could use in saving fuel.
2. I'm not sure whether I'll use styrofoam or vermiculite to insulate the exterior walls.

Capitalize the names of persons, races, and nationalities:

| Alexis | Maxwell R. Folger | Cuban | African |
| American | Mrs. Jameson | Chicano | Chinese |

Black (Afro-American) is usually not capitalized either as a noun or as an adjective.

Capitalize place names:

| Shreveport | Puerto Rico | South Carolina | Lake Mead |
| High Point | Zaire | Polk County | Columbia River |

Capitalize the names of organizations and historical events:

| ACLU | General Motors | Southeast Asian Conflict |
| World War I | IBM | the Treaty of Paris |

Capitalize the days of the week, months, and holidays:

Easter   July   Tuesday   Yom Kippur   Thanksgiving

Capitalize titles preceding someone's name:

| Dr. Taylor | Chairman Ghent | Assistant Chief Harkins |
| Chancellor Richbury | Colonel Samples | Professor Chittendon |

Similar titles used after a name are usually not capitalized:

Wendel Cheatam, president of First National Bank
Marlene Winbush, city judge

Capitalize common nouns such as *street, river,* and *company* only when they are part of a proper noun:

Devex Corporation   Arden Street   American Airlines   Uncle Bill
Shimer College   Pacific Ocean   Psychology 381   Glumburg High School

Capitalize *north, south, east,* and *west* only when they refer to specific places:

the Midwest   South Georgia   the West Coast

Capitalize the first word and all important words (omit *a, an, the,* conjunctions, and short prepositions) of the title of a book, periodical, article, report, or other major document. Capitalize also the titles of chapters or other major divisions of such a work:

The Pilot   *Scientific American*   "Methodology Used"
"Proposed Changes"   *Florida Plants*   *Datamation*

Apostrophes

Apostrophes serve two main functions: they show possession in nouns and certain pronouns; and they make contractions. They are also used in some plurals, but less frequently now than they formerly were.

A noun or indefinite pronoun is made possessive by adding either an apostrophe or an apostrophe and an *s*. If the word does not end in *s*, add an apostrophe and an *s:*

lady's desk   men's rest room   women's lounge
children's books   man's pen   Joanne's calculator

If a singular word ends in *s*, add apostrophe *s* unless the result would be difficult to pronounce:

Mavis's tools   James's books   Aristophanes' plays   Onassis' will

If a plural ends in *s*, add only an apostrophe:

ladies' lounge   doctors' quarters

In compound nouns, use the apostrophe only with the last word:

brother-in-law's (singular possessive)   passer-by's (singular)
brothers-in-law's (plural possessive)   passers-by's (plural)

Apostrophes are also used to form contractions:

don't   aren't   they'll   doesn't   I'm

Note that *its,* the possessive of *it,* has no apostrophe, while *it's,* the contraction of *it is,* does have one.

Common contractions are now accepted in nearly all types of writing; however, in highly formal writing you would be safer to avoid them.

Apostrophes are sometimes used with the plurals of letters, numbers, and words used as words. The words, letters, and numbers are also italicized:

1. The good writer is careful not to use too many *and's* or *but's.*
2. The *h's* are silent in *honor* and *honest.*
3. I do not remember the entire serial number, but it started with several *7's.*

In each case, especially with numbers, the plural could be correctly formed without the apostrophe.

Abbreviations

Determining when to abbreviate is often difficult. A few abbreviations, such as A.M., *Ms., Mr., Dr.,* and *NASA,* are used regularly in all forms of writing. Highly formal writing excludes almost all other abbreviations. Illustrations and very compact written forms such as nurses' notes require the use of all possible abbreviations. Most job-related writing falls between the two extremes. The following guidelines should help:

1. Never use an abbreviation your intended reader might not understand.
2. In extremely informal writing use abbreviations as you see fit, depending on your personal knowledge of the reader.
3. In semiformal and highly formal writing, avoid abbreviations in the text as much as possible. Use abbreviations to save space in tables, figures, specifications, parts or price lists, and footnotes.
4. When a term to be abbreviated is to be used several times in the same text, spell it out the first time with the abbreviation in parentheses.
5. Avoid signs such as " for inches and # for pounds except in illustrations.
6. When in doubt, spell it out.
7. Acronyms (abbreviations pronounced as words) are capitalized but not punctuated:

NATO   North Atlantic Treaty Organization
UNESCO United Nations Educational, Scientific, and Cultural
        Organization

8.   Abbreviations for names of companies, governmental agencies, and other organizations are generally capitalized and written without punctuation:

FHA   Federal Housing Authority
CPS   Certified Professional Secretary
GAC   General Acceptance Corporation
CIA   Central Intelligence Agency

9.   Lowercase abbreviations are generally separated and followed by periods, although the periods are often omitted with units of measurement:

f.o.b.   c.o.d.   ft. *or* ft   h.p. *or* hp

10.   Abbreviate titles such as *doctor, captain, reverend,* and *lieutenant* only when they precede a proper name:

Ens. Robert Thomas   Rev. Lionel Willows   Dr. Theodore Dippy

11.   Use documentary abbreviations such as *ibid., col.* (column), *rev., p.,* or *pp.* as needed in footnotes or reference lists, but avoid them in the main text.

12.   Spell out place names unless you are certain no possible confusion could result from abbreviating them.

13.   Following are lists (arranged by type) of common abbreviations:

Titles

Atty.	Attorney	M.B.A.	Master of Business
B.A.	Bachelor of Arts		Administration
B.S.	Bachelor of Science	M.D.	Doctor of Medicine
D.D.S.	Doctor of Dental Science	Ph.D.	Doctor of Philosophy
Dr.	Doctor (any doctor's degree)	CPA	Certified Public Accountant
Hon.	Honorable	CPS	Certified Professional Secretary
M.A.	Master of Arts	CDP	Certified Data Processor

Organization Types

Assn.	Association	Corp.	Corporation
Bros.	Brothers	Inc.	Incorporated
Co.	Company	Ltd.	Limited

Weights, Measures, and Other Units

ac	alternating current	F	Fahrenheit
AM	amplitude modulation	FM	frequency modulation
amp	ampere	ft	foot
bbl	barrel	g	gram
Btu	British thermal unit	gal	gallon
bu	bushel	hp	horsepower
C	centigrade or Celsius	hr or h	hour
cal	calorie	Hz	hertz

cc	cubic centimeter		in	inch
cm	centimeter		kg	kilogram
cps	cycles per second		km	kilometer
cu ft	cubic foot		kph	kilometers per hour
cu in	cubic inch		kv	kilovolt
db	decibel		kw	kilowatt
dc	direct current		kwh	kilowatt hour
doz	dozen		l	liter
lb	pound		qt	quart
m	meter		rpm	revolutions per minute
mg	milligram		sec	second
min	minute		sq	square
ml	milliliter		t	ton
mm	millimeter		v	volt
mo	month		w	watt
mph	miles per hour		wk	week
oz	ounce		yd	yard
psi	pounds per square inch		yr	year

Footnotes, Bibliographies, and Manuscripts

app.	appendix		MS(s)	manuscript(s)
ca.	about (a specific time)		n.	note
cf.	compare		n.d.	no date
ch.	chapter		no(s).	number(s)
col.	column		n.p.	no place (of publication)
ed.	editor or edited by		p(p).	page(s)
f. or ff.	the following page or pages		rev.	revised
ibid.	in the same place		sec.	section
i.e.	that is		trans.	translated by
il.	illustrated		vol(s).	volume(s)

State Abbreviations Recommended by U.S. Post Office

Alabama	AL		Massachusetts	MA
Alaska	AK		Michigan	MI
Arizona	AZ		Minnesota	MN
Arkansas	AR		Mississippi	MS
California	CA		Missouri	MO
Colorado	CO		Montana	MT
Connecticut	CT		Nebraska	NE
Delaware	DE		Nevada	NV
District of Columbia	DC		New Hampshire	NH
Florida	FL		New Jersey	NJ
Georgia	GA		New Mexico	NM
Hawaii	HI		New York	NY
Idaho	ID		North Carolina	NC
Illinois	IL		North Dakota	ND
Indiana	IN		Ohio	OH
Iowa	IA		Oklahoma	OK

Kansas	KS	Oregon	OR
Kentucky	KY	Pennsylvania	PA
Louisiana	LA	Puerto Rico	PR
Maine	ME	Rhode Island	RI
Maryland	MD	South Carolina	SC
South Dakota	SD	Virginia	VA
Tennessee	TN	Washington	WA
Texas	TX	West Virginia	WV
Utah	UT	Wisconsin	WI
Vermont	VT	Wyoming	WY

Using Numbers

Authorities disagree about which numbers can be written as numerals and which should be written out as words. A good safe practice, though, is to write out one-and two-word numbers and to use numerals for all others.

seventeen	nine	three million	1,319
twenty-six	206	one hundred	5,687,414

Certain numbers are almost always expressed in numerals: page numbers, units of measure, percentages, decimals, fractions with whole numbers, building and highway numbers, time when followed by A.M. or P.M., and dates.

page 23	14 miles	220 volts	0°C
9661 East Robs St.	63.21 percent	U.S. 51	Illinois 47
12:06 P.M.	7.3 kilos	31¼ inches	14.68 centimeters

Certain other numbers are almost always written out: numbers beginning a sentence, approximations, numbered streets (unless space does not permit), and time not followed by A.M. or P.M.

six-thirty	almost four hundred	East Fifty-first Street
seven o'clock	around three hundred fifty	1200 West Twelfth Street

One hundred eighteen visitors attended our open house.

Proper expression of numbers often varies from one profession or even one company to another, so always check to be sure.

Manuscript Form

Business and technical writing varies so much from profession to profession and from organization to organization that generalizations about manuscript form are risky. Many organizations and professions print specific style manuals governing the form of materials submitted; others have guidelines less formal but no less stringent. Additionally, different types of job-related documents are conventionally done in different forms. This text has included suggestions on manuscript form in the sections dealing with specific types of writing. Perhaps the most valuable suggestion it can offer here is that you should always check on any restrictions or guidelines for manuscript form before composing a new type of communication or before preparing a communication for a new organization.

The following suggestions pertain to preparing manuscripts in general. In the absence of more concrete company or professional guidelines, you can rely on them to reflect conventional practice:

1. *Paper:* Always use standard 8½- by 11-inch paper unless instructed to do otherwise or unless the document in question is generally done on paper of another size. Use good-quality paper, avoiding onionskin except for carbon copies. Even then, most readers prefer a photocopy to an onionskin carbon.

2. *Typing and writing longhand:* Type as much of your written communication as possible. Typing is both more legible and more impressive than longhand. Formal documents and almost any document sent to an outside agency should always be typed. If you must submit some college papers or perhaps some informal business papers in longhand, strive for legibility first, then for pleasing appearance. Do not use closely lined paper or sheets torn from spiral notebooks. Recopy for clarity and appearance unless you are writing a very casual note to a colleague.

3. *Margins and spacing:* Double-space college papers and formal reports and proposals. Most other material can be single-spaced. An inch-and-one-half margin on the left side and top, with an inch at the right and bottom, is standard. Single-space footnotes and long quotations. Indent all paragraphs five spaces. Leave an extra space between any heading or title and the text that follows it.

4. *Word division:* Only if you are specifically trying to leave your right margin perfectly even will you have to divide many words at the ends of lines. When you must divide a word, follow these suggestions: Never divide a one-syllable word. Divide other words only between syllables, never leaving a one-letter syllable to stand alone at the beginning or end of a line. Avoid dividing a prefix or suffix from its root word; divide hyphenated words only at the hyphen. Never divide numbers expressed in numerals.

5. *Pagination:* Front matter (anything preceding the main text) is generally numbered in lowercase Roman numerals (*i, ii, iii, iv*) at the bottom center. Everything else is numbered in ordinary Arabic numerals (*1, 2, 3, 4*) in the upper right-hand corner. Number consecutively through the end of the document, including back matter such as appendixes and bibliographies. Omit the number from page 1 but put *2* on the next page.

18

Research Papers

Overview   *The library research paper is very much like a long report, but it does have special features. Find and narrow a topic before beginning research. While collecting material, be careful to keep sources straight, to distinguish between quoted and paraphrased material, and to record all necessary bibliographical information. Learn when and how to give references.*

One of the most common types of writing required in the college classroom is the library paper (term paper, research paper). This paper is much like the long report explained in Chapter 10, but it has certain important differences. These differences are the focus of this section.

The Process

The process of putting together a good library research paper can be divided into the following ten steps:

1. Select a general subject.
2. Do some preliminary reading.
3. Narrow the subject and formulate a thesis or purpose statement.
4. Prepare a rough outline.
5. Prepare a working bibliography.
6. Take notes.
7. Develop a detailed outline.
8. Write a rough draft.
9. Revise and polish.
10. Make a final copy for submission.

1. *Select a general subject:* If no subject has been specifically assigned, study any guidelines or suggested topics given. Look for a subject that you can find information about. It will do you no good to come up with an exciting but obscure topic about which you can find only two magazine articles. You need to find enough different sources of information that you can judge each one on the basis of the others and arrive at your own conclusions. So consider the kinds of material available in your library. Pick a topic fresh enough that you can say something worthwhile about it; stay away from old saws such as euthanasia, abortion, or marijuana. Avoid cut-and-dried subjects such as "The Life of Benjamin Franklin" unless you think you can come up with something original. Finally, look for a subject that interests you. That way your work will be much more enjoyable.

2. *Do some preliminary reading:* Get yourself acquainted with the general area. Browse through a reference book, a chapter from a textbook, or even a few encyclopedia articles. Discover the parts and nuances of the subject.

You are not looking for specific data yet, just getting a working knowledge of the subject.

3. *Narrow the subject and formulate a thesis or purpose statement:* Now try to limit your subject to something workable. Try to arrive at a topic you can reasonably expect to cover fully but not something ridiculously minute. (Do not be surprised if you have to narrow it again after you have done more research.) Formulate a question that you expect your research to answer or a purpose that your paper should fulfill. Here you can apply the preliminary questions discussed in Chapter 1. A general idea about your topic ("I am going to discuss some of the new chemicals used to develop and print photographs") will not give you the kind of focus you need. Try for a more specific purpose: "I intend to show amateur photographers that new film and new chemicals make it easy to work with color in the home darkroom."

4. *Prepare a rough outline:* Determine what seem to be the three or four most logical divisions of your proposed paper. If you can further subdivide some of these divisions, do so. Do not worry if you have to change your outline as you gather material. The purpose of this rough, working outline is to help you to identify relevant material from your sources.

5. *Prepare a working bibliography:* Now the actual research begins. Start by putting together a list of periodical articles, books, and other materials relevant to your outline. Your primary sources for this list will be the card catalogue, the *Readers' Guide to Periodical Literature,* and specialized references. Prepare a bibliography card for each possible source, following the guidelines presented later in this chapter.

6. *Take notes:* Now go to book stacks (or to the circulation desk if your library has closed stacks) and start locating the books and articles on your list. As you locate a potential source of information, browse through it carefully but quickly. Check tables of contents and indexes when available. Look over a preface, introduction, or abstract if one is available; skim the headings and the boldface print. With a little practice you can learn to determine in two or three minutes whether a given article or even a book has any information you can use. When you do find useful information, read through it and take down its important points on note cards.

7. *Develop a detailed outline:* Armed with a healthy stack of note cards, you are ready to begin putting your paper together. Read over your cards and sort them according to your rough outline. Study each set to make certain that you have all the information you need. Then develop a detailed outline by dividing and subdividing each point.

8. *Write a rough draft:* Using your detailed outline, begin composing the paper. Use the suggestions given in Chapter 5. Be particularly careful to mark spots where you have borrowed information so that you can put in reference numbers or footnotes.

9. *Revise and polish:* Using the suggestions in Chapter 6, edit your draft, making a second rough draft if major changes seem necessary. Otherwise, make all changes right on the rough draft. Write up your detailed outline in proper form; prepare a title page similar to those in Chapter 10; write up a complete bibliography or reference list; prepare footnotes if needed; proofread the entire paper.

10.   *Make a final copy for submission:* All that remains is to type up (or write longhand if your instructor permits) a polished copy for submission. Give it a final proofreading, making last-minute corrections unobtrusively in dark ink. Put it into a sturdy cover. Submit it.

Finding Information in the Library

The first and most important places to look for information are the card catalog and the indexes to periodicals. Then you can check reference books and the vertical file.

Card Catalog

In most libraries every book and audio-visual aid is indexed in a card catalog. Most card catalogs list works by author, title, and subject, thus having three cards for each item. Some libraries separate the three sections; others combine them; in either case, cards are filed alphabetically.

When you find a card of a promising source, note the call number in the upper left-hand corner and the author's name and the title on the first two lines. You'll need this information either to find the book yourself or to put on a call slip if the book stacks are closed to students.

Periodical Indexes

Organized alphabetically by subject, periodical indexes list articles, giving their authors' names, their titles, and the names and dates of the journals they appeared in. Some guides include brief abstracts of the articles. Best known is the *Reader's Guide to Periodical Literature,* which indexes articles from a wide variety of popular magazines. Other indexes list articles in specialized subject areas. Here are some of the more important ones.

Agriculture Index
Applied Science and Technology Index
Business Periodicals Index
Chemical Abstracts
Cumulative Index to Nursing Literature
Education Index
Engineering Index
International Nursing Index
New York Times Index
Psychological Abstracts
Social Sciences and Humanities Index
U.S. Government Publications

Check your library reference section to see if it has any other periodical guides in your subject area. Once you have located a relevant periodical guide, you can make a list of potentially useful articles, then check your library's periodical holdings list to see which articles are available. Only the very largest research libraries subscribe to all the many journals published; however, if an especially promising source is unavailable, the librarian may be able to get it for you on an interlibrary loan.

Reference Books

The reference section of the library is usually located on the main floor near the card catalog and the periodical guides. Here you will find a variety of specialized volumes ideal for locating specific pieces of information and for doing preliminary reading. Almanacs giving general statistical information are included, along with many specialized handbooks. You will find dictionaries of all sorts, from unabridged, general dictionaries to specialized ones, including biographical dictionaries listing prominent figures in most areas. Encyclopedias range from the common general-purpose sets to individual volumes on specific topics. Finally, bibliographies list available information on many subjects.

One relatively new source is the computer reference bank. Many libraries are now connected with computer services offering a wide range of information.

Vertical File

Usually located near the reference section, the vertical file contains drawers or large folders of miscellaneous information on a wide range of subjects. Each file, listed alphabetically, contains brochures, pamphlets, newsletters, booklets, and selected newspaper clippings. Check the vertical file, especially if you are researching a popular subject.

Special Features

Using Other People's Material

The most common problem faced by the inexperienced research-paper writer is determining how to fit together large masses of material from various outside sources. Far too often, the resulting paper is nothing more than a mosaic, splicing together pieces of data from various sources with nothing of the writer's own involved.

One way to avoid this problem is to follow the process explained earlier. Examine the subject from all sides. Read the explanations or descriptions of several authors. Become knowledgeable—if not actually authoritative—on your subject. Formulate your own ideas based on those of others. When you have developed your own thesis, make sure that all the material you use in your paper helps to establish that thesis.

What to Footnote or Reference

Another common problem is determining where to put footnote or reference numbers. Failure to use enough of them can rob the reader of important information, and it can cause inadvertent plagiarism. On the other hand, unnecessary reference numbers distract and perhaps offend the reader—not to mention wasting a great deal of your time.

Document any direct quotation. Document any information—even if it is expressed in your own words—that can be directly attributed to a specific source. You need not document material found in several sources if it is obviously basic general knowledge in the field. You need not document information you already knew if that information can be considered common knowledge. But be careful about dismissing information as common knowledge. When in doubt, document.

Put footnote or reference numbers one-half space above the line at the end of a sentence, even if the borrowed material actually occurs earlier in the sentence. For lengthy sections of material borrowed from the same source, use a number after the final sentence of each paragraph. But beware of such long quotations: they may indicate that you are relying too heavily on one source instead of developing your own ideas.

Taking Good Notes

Many inexperienced research paper writers spend long hours working in the library, accumulating large masses of notes. Later they find they cannot decipher these notes to compose the paper. Thoughtful notetaking can minimize if not totally eliminate this problem.

Use note cards, not just a section in a notebook (4 by 6 cards work nicely). Put only one note on a card. If you need a number of cards for notes from one source, you may waste a few cards, but you will save time and confusion. Code each note card in two ways: indicate the source and page of the information, and indicate the section in your rough outline relevant to the note. Thus, when you begin writing your paper, you will know what to footnote, and you will be able to sort your note cards into stacks pertaining to each section of the outline.

Use your own words for the note card. Should you copy directly from the source intending to change to your words in the finished paper, you can easily forget and inadvertently plagiarize. Make the notes thorough enough that you can understand them later. Scrawled, abbreviated notes that are perfectly clear today will not necessarily be so clear three weeks from now. Material you may want to quote directly in the paper should be quoted directly on the note cards, with quotation marks used as a reminder. Do not copy down everything in the article or book. Jot down key points, keeping your thesis in mind. Following are two sample note cards:

Refers to Part III B
of rough outline

Refers to
bibliography

$\underline{III}\ B$ 3- pp. 21-22

      Initiating new graphic sheet is listed as task #48. Sheet must be neat and legible, photocopyable, in the appropriate color ink.

      No ink eradication or erasing allowed.

      Very thorough and rigid

Refers to Part II
of outline

Same source,
different pages

$\underline{II}$ 3- p. 91

      Preparing and routing special therapy requisitions involves the following steps:

1) Interpret & transcribe orders, etc.
2) Use addressograph to imprint requisitions.
3) Complete requisitions.
4) Get nurse's initials.
5) Route, file, & enter on computer.

CRM: "Using material provided, prepare and route special therapy requisitions on a 'liver scan.'"

Preparing Bibliography Cards

Neglecting to prepare bibliography cards or preparing them carelessly can cause a dilemma when you try to put the finished paper together. You will find material that you want to use and no source given for it, or you will find that you have most of the needed bibliographic information but are missing a page number, a publisher's name, or the like. Then you will either have to trek back to the library or risk using an incomplete entry. So resist the temptation to scribble down sources on a sheet of paper; use individual note cards for each.

A good bibliography card contains all necessary information in correct form for writing individual bibliography entries. Footnotes can also be written quite easily by using the information on the appropriate bibliography card and the page numbers from the note card. Next are two sample bibliography cards. Further samples of bibliography form are found on page 382.

Effective Documentation

College instructors frequently specify a particular style for documentation and recommend particular style sheets or manuals such as *A Manual of Style*, 12th ed. (Chicago: University of Chicago Press, 1969); the *MLA Style Sheet*, 2d ed. (New York: Modern Language Association, 1970); or James D. Lester, *Writing Research Papers*, 3d ed. (Glenview, Ill.: Scott, Foresman and Company, 1980). Many professional associations have also developed style sheets for use in for-

**Number in preliminary
bibliography**

5

*Mitchell, John H. "Teaching Technical
Writing and Business Communications
in Australia."*

*The Technical Writing Teacher, 2
(Fall 1976), 12-19.*

> 2
>
> *Coffin, Kenneth B. and Louis A. Leslie.*
> *Handbook for the Legal Secretary.*
> *New York: McGraw-Hill Book*
> *Company, 1968.*

mal papers within their disciplines. Always check to determine if a particular style is required or recommended. If so, you can usually find an inexpensive manual in the college bookstore or in the library. Use it faithfully.

If no particular style is required or recommended, choose one that you feel comfortable using. The styles used in this text are both popular and relatively simple. Two approaches are discussed: the reference list and footnotes with bibliography. Traditionally, footnotes and bibliography have been far more common and are still preferred by most style manuals and college instructors. However, many disciplines are now turning to the somewhat simpler reference list, and it is rapidly gaining acceptance.

Reference List

To document using a reference list, first develop a numbered list of all sources used in the paper. Then use numbers in the text (often termed *call-outs*) to refer to the numbered sources on the list. A colon separates the call-out number from a second number or pair of numbers indicating the exact pages referred to (3: 56–60).

Entries in the reference list are sometimes put in alphabetical order and sometimes in order of call (the sequence in which they are first referred to in the text of the paper). You may repeat the same call-out number many times as you refer again to the same source. If you have six sources in your reference list, your call-outs will be the numbers *1* through *6* appearing perhaps several times each. Sample call-outs and reference list follow.

The format for the entries in the reference list is the same as that shown for a bibliography—unless, of course, a particular style is required by your instructor, organization, or profession.

. . . The Red Cross defines a wound as "a break in the skin or mucous membrane."$^{(1:11)}$

. . . The best way to stop bleeding of small wounds is to press on them with a piece of sterile gauze.$^{(2:220)}$

. . . An infection in a wound will show up as early as two days later or may take as long as a week.$^{(1:13)}$

References

1. American National Red Cross. First Aid Textbook. 4th ed., rev. Garden City, New York: Doubleday and Company, 1966.

2. Fishbein, Morris. The Handy Home Medical Advisor. Garden City, New York: Doubleday and Company, 1963.

Footnotes and Bibliography

The standard form of documentation historically has comprised footnotes clearly identifying the source of each call-out number used in the text and an alphabetized bibliography following the text. Footnotes formerly appeared at the bottom of the page where the call-out appeared. Now they are frequently placed on a footnote page following the text but preceding the bibliography. If your style manual requires footnotes and bibliography, check to determine where to put the footnotes.

Call-out numbers for footnotes—like those for a reference list—always come at the end of the sentence containing the relevant material. However, they are not followed by specific page numbers, as this information is given in the footnote. They also differ from reference list call-outs in that they are consecutively numbered. Thus, if you referred to only six sources but referred to some several times for a total of thirteen call-outs, you would have thirteen footnotes numbered *1* through *13*.

Footnotes are likely to take two forms: The first time a source is referred to, a complete footnote is used. Subsequent references to that source need not contain full citations. Current practice often prefers that they contain the author's last name and the appropriate page number. However, when two successive footnotes refer to the same source and same page, the second may consist of *Ibid.*; and when two successive footnotes refer to the same source but to different pages, the second may consist of *Ibid.* followed by the appropriate page number.

The bibliography is an alphabetical list of your sources, giving complete publication information about each. Whereas the footnotes show readers exactly where you found each piece of information, the bibliography shows them the complete details about each source so that they can locate the information themselves.

Three types of bibliographies are found: full bibliographies contain all sources used in the paper; selected bibliographies list only sources actually cited in footnotes; annotated bibliographies contain comments about the specific information in or usefulness of the works listed. Sample footnotes and a sample bibliography follow.

Sample Footnotes

Book:

¹Edwin Rams, _Analysis and Variation of Retail Locations_ (Englewood Cliffs, N.J.: Prentice-Hall, 1976), pp. 76-78.

Book with more than one author:

²Sarah E. Archer and Ruth Fleshman, _Community Health Nursing_ (N. Scituate, Mass.: Duxbury Press, 1970), pp. 8-9.

Edition other than the first:

³Marshall H. Jurgens, _Applied Animal Feeding and Nutrition_, 3rd ed. (Dubuque, Iowa: Kendall/Hunt, 1974), p. 108.

Book compiled by an editor:

⁴Thomas F. Adams, ed., _Criminal Justice: Readings_ (Pacific Palisades, Calif.: Goodyear, 1971), pp. 225-36.

Newspaper article:

⁵"Holiday Death Rate Declines," _San Francisco Chronicle_, March 17, 1972, Section B, p. 2.

Magazine article:

⁶Walter Litten, "The Most Poisonous Mushrooms," _Scientific American_, March 1975, pp. 90-91.

Article in a professional journal:

⁷Abraham Ellenbogen, "An Experiment in Teaching Typing to Younger Students," _Business Education Forum_, 23 (1968), 13-14.

Encyclopedia article:

⁸Donald M. O'Brien, "Fire Fighting and Prevention," _Encyclopedia Americana_, 1973.

Sample Bibliography

The following bibliography shows the forms for the sources shown in the sample footnotes. Note the differences so that you do not confuse the two forms.

Adams, Thomas F., ed. Criminal Justice: Readings. Pacific Palisades, Calif.: Goodyear, 1971.

Archer, Sarah E., and Ruth Fleshman. Community Health Nursing. N. Scituate, Mass.: Duxbury Press, 1970.

Ellenbogen, Abraham. "An Experiment in Teaching Typing to Younger Students." Business Education Forum, 23 (1968), 13-26.

"Holiday Death Rate Declines." San Francisco Chronicle, March 17, 1972, Section B, p. 2.

Jurgens, Marshall H. Applied Animal Feeding and Nutrition. 3rd ed. Dubuque, Iowa: Kendall/Hunt, 1974.

Litten, Walter. "The Most Poisonous Mushrooms." Scientific American, March 1975, pp. 90-101.

O'Brien, Donald M. "Fire Fighting and Prevention." Encyclopedia Americana. New York: Americana Corporation, 1973.

Rams, Edwin. Analysis and Variation of Retail Locations. Englewood Cliffs, N.J.: Prentice-Hall, 1976.

Index